D1037820

X-EVENTS

X-EVENTS

THE COLLAPSE OF EVERYTHING

JOHN CASTI

WILLIAM MORROW

An Imprint of HarperCollins*Publishers*

To connoisseurs of the unknown unknowns

X-EVENTS. Copyright © 2012 by John L. Casti. All rights reserved. Printed in the United States of America. No part of this book may be used or reproduced in any manner whatsoever without written permission except in the case of brief quotations embodied in critical articles and reviews. For information address HarperCollins Publishers, 10 East 53rd Street, New York, NY 10022.

HarperCollins books may be purchased for educational, business, or sales promotional use. For information please write: Special Markets Department, HarperCollins Publishers, 10 East 53rd Street, New York, NY 10022.

A hardcover edition of this book was published in 2012 by William Morrow, an imprint of HarperCollins Publishers.

FIRST WILLIAM MORROW PAPERBACK EDITION PUBLISHED 2013.

Designed by Jamie Lynn Kerner

Library of Congress Cataloging-in-Publication Data has been applied for.

ISBN 978-0-06-208829-1

13 14 15 16 17 OV/RRD 10 9 8 7 6 5 4 3 2 1

CONTENTS

AUTHOR'S NOTE / VII

PREAMBLE: PUTTING THE "X" INTO X-EVENTS / 1

PART I: WHY NORMAL ISN'T SO "NORMAL" ANYMORE / 17

PART II: GETTING DOWN TO CASES / 59

X-EVENT 1: DIGITAL DARKNESS / 68
A Long-Term, Widespread Failure of the Internet

X-EVENT 2: WHEN DO WE EAT? / 92
Breakdown of the Global Food-Supply System

X-EVENT 3: THE DAY THE ELECTRONICS DIED / 109
A Continent-Wide Electromagnetic Pulse Destroys All Electronics

X-EVENT 4: A NEW WORLD DISORDER / 122
The Collapse of Globalization

X-EVENT 5: DEATH BY PHYSICS / 144
Destruction of the Earth Through the Creation of Exotic Particles

X-EVENT 6: BLOWN AWAY / 161
Destabilization of the Nuclear Landscape

X-EVENT 7: RUNNING ON EMPTY / 180
Drying Up of World Oil Supplies

X-EVENT 8: I'M SICK OF IT / **195**
A Global Pandemic

X-EVENT 9: DARK AND DRY / **212**
Failure of the Electric Power Grid and Clean Water Supply

X-EVENT 10: TECHNOLOGY RUN AMOK / **233**
Intelligent Robots Overthrow Humanity

X-EVENT 11: THE GREAT UNWINDING / **251**
Global Deflation and the Collapse of World Financial Markets

PART III: X-EVENTS REDUX / **271**

NOTES AND REFERENCES / **303**
INDEX / **318**

AUTHOR'S NOTE

GLANCING QUICKLY AT THE COVER OF THIS BOOK, IT WOULD BE EASY to think it is yet one more tale of doom and gloom, an account of apocalypses waiting to jump out at humanity and send our way of life tumbling back to a preindustrial level. But as often turns out to be the case in life, first impressions can be deceiving, if not just flat-out wrong, and this is not a volume aimed at scaring the bejesus out of you. Quite the opposite, in fact. So if it's not a vision of hellish things to come, what, pray tell, is it?

The book you hold in your hands is a tale of *possibilities,* dramatic possibilities that are rare, surprising, have potentially huge impact on human life, and about which we cling to the illusion that they are not "made in my backyard." In the vernacular, these possibilities are generally lumped into the catchall term "extreme events." I like *X-events* better. This is a book about these outlier events, those surprises that are complementary to everything that takes place in what we will call the "normal" region. By way of contrast, the X-events region is one that has been far less scientifically investigated, just because its elements, ranging from asteroid impacts to financial market meltdowns to nuclear attacks, are by their very nature rare and surprising. Science is mostly about the study of repeatable phenomena; X-events fall outside that category, which is a major reason why at present we have no decent theory for when, how, and why they occur. If nothing else, this book is a call to arms for the development of that theory, what we might label "a theory of surprise." We might compactly summarize the situation that the book's primary goal is to propose an answer to the question: How do we characterize risk in situations where probability theory and statistics cannot be employed?

X-events of the human—rather than nature-caused—variety are the result of too little understanding chasing too much complexity in our human systems. The X-event, be it a political revolution, a crash of the Internet, or the collapse of a civilization, is human nature's way of reducing a complexity overload that has become unsustainable. Each part of the book aims at shedding light on the following questions:

- Why do X-events occur?
- Why are X-events happening more frequently nowadays than ever before?
- What impact will a particular X-event have on our twenty-first-century lifestyle?
- How can we anticipate when the risk of an X-event taking place has risen to a dangerous level?
- When can we do something to prevent an impending X-event and when can we only prepare to survive it?

The answer to each of these questions is in its own way tied up with the exponentially increasing levels of complexity necessary to preserve the critical infrastructures of modern life. That fact is the thread running through this book.

X-Events is a work of concepts and ideas. To make these pages as accessible as possible to a nontechnical reader, I've used no formulas, charts, equations, graphs, or jargon. (Well, one chart!) Fundamentally, the book is a collection of linked stories that together make the case that complexity can—and will—kill if we let it get out of control.

As is so often the case, hindsight is 20/20. And so it is with this book, too. After completing the draft manuscript, it dawned on me that the book you hold is actually the second volume of an emerging trilogy on human social events, their causes, and their consequences.

The first volume was my 2010 work *Mood Matters*, which addressed the social psychology of groups and how that "social mood" biases the sorts of collective events we can expect to see on all timescales. *X-Events* focuses on the root cause of human-generated extreme events, and the manner in which such events can change our way of life. The third volume will outline how extreme events are as much an opportunity as a problem, the "creative" half of economist Joseph Schumpeter's famous notion of "creative destruction."

Readers who wish to send me their comments, ideas, and/or complaints, can do so via the web address: john@moodmatters.net.

THE MOST SATISFYING PART ABOUT COMPLETING ANY BOOK IS THE OPportunity to thank the many people who helped in its creation. In this regard, I'm more blessed than most by having had the benefit of many "beta" readers, who tirelessly and generously gave their time to making this a better book than I had any right to expect. So I'm happy to acknowledge their efforts in print and thank them publicly for their generosity and good judgment. So in no particular order, kudos to Olav Ruud, Brian Fath, Leena Ilmola, Jo-Ann Polise, Helmut Kroiss, Rex Cumming, Adam Dixon, and Timo Hämäläinen for their thoughts, suggestions, and contributions to one or more of the book's chapters. A special tip of the hat to my most faithful readers, Trudy Draper and Zac Bharucha, who read every line of every chapter and did their absolute best to look after the reader's interests. Whatever incomprehensibilities remain in the text are there despite their best efforts to get me to fix them. Finally, my "tough love" publisher, Peter Hubbard, senior editor at William Morrow/HarperCollins, forced me to write and rewrite passage after passage to get it right. Without his enthusiasm and continuing support, this book would never have seen the light of day.

John Casti
Vienna, Austria
November 2011

PREAMBLE

PUTTING THE "X" INTO X-EVENTS

THE COMPLEXITY TRAP

IN EARLY 2010, AMERICAN ARCHITECT BRYAN BERG COMPLETED what is still the world's largest house of cards. Using more than four thousand decks, he constructed a towering replica of the Venetian Macao-Resort-Hotel in China that stands over ten feet high and is thirty feet wide. Viewing this amazing structure, I immediately saw it as a kind of metaphor for the highly complex, interconnected world we live in today. A mouse scurrying about in the room or a sneeze by a visitor at the wrong moment, and the forty-four days Berg spent building his house of cards could be blown away in a second. So, too, with the ultrafragile infrastructures we depend on for everyday life.

The entire industrialized world relies on a continuing infusion of ever more advanced technology. Moreover, the systems that underwrite our lifestyle are completely intertwined: the Internet depends on the electrical power grid, which in turn relies on energy supply from oil, coal, and nuclear fission, which likewise rests on manufacturing technologies that themselves require electricity. And so it goes, one system piled upon another piled upon yet another, everything connected to everything else. Modern society is just like the cards in "Berg's casino," where each new card sits atop those below it. Such a setup is an open invitation to that mouse scurrying across the floor to nudge a low-level card and send the entire structure tumbling down.

Of course, the fragility of the construction is what gives a house of cards its cachet. This is great—for a parlor trick. But do we want to have our entire way of life rest upon a house of cards? Imagine New York or Paris or Moscow without electricity for an uncertain period of time. Or to take a longer-term view, what if we have no new technology for a decade? What happens then to our standard of living?

This is a great question. What does happen to our standard of living when the seductive music of technology stops? Even more interesting, what could stop the music? Like all fundamental questions, this one too has many surface answers. But all these replies rest upon a more foundational reason for why and how technology can stop. In these pages, I argue that the deep reason for how the music stops is that it's the game changer, the extreme event, that pulls the plug. And these high-surprise, high-impact "X-events" that unravel systems are themselves driven by the ever-increasing complexity of technological and other human-created infrastructures, the very infrastructures that sustain what might euphemistically be termed "normal" life. Part of the story I tell here is to graphically point out that this so-called normalcy has been acquired at the huge cost of high vulnerability to collapse at the hands of an increasingly broad spectrum of X-events. Moreover, every one of these possible game changers has the same root cause: far too little understanding of the wondrous and counterintuitive ways of complex systems.

I've spent most of my professional life exploring complexity while working at organizations like the RAND Corporation, Santa Fe Institute, and International Institute for Applied Systems Analysis (IIASA). When I got my Ph.D. in mathematics in the 1970s and began investigating complex systems, the world was a very different place. Phones had rotary dials, computers cost millions of dollars, half the world was closed to free trade and travel, and you could actually fix your old Chevy or VW without a degree in electrical engineering. Indeed, you don't have to be a system theorist to see that our lives and societies have become ever more dependent on ever more opaque technology. A large part of this dependence rests on the increased complexity of the technologies themselves. As each year goes by, the complexity of our devices and infrastructures ranging from automobiles to finance to power grids to food-supply chains grows exponentially. A part of this increase is intended to buy a level of robustness and protection against system failure, but usually just in the face of minor, fairly predictable types of shocks. But most of it doesn't. Do you really need an espresso machine with a microprocessor? Must we have seventeen varieties of dog food on offer at the supermarket? Is it necessary to make cars requiring an inches-thick owner's manual to explain how the power seats, the GPS system, and other gadgetry built in to the car function?

These little examples of everyday complexity increase are often packaged as technological success stories. But are they really? There's a strong case to be made that they are in fact technological failures, not successes at all, when you add up the time spent analyzing the ingredients of the dog food on offer before making a choice that's more illusory than real or factor in the frustration you experience thumbing through that owner's manual looking for that key page telling you how to set the clock on your new car. But unwanted/unneeded features in your new car or distinctions without a difference at the food store are just minor, laughable even, annoyances. (Un)luckily, you don't have to go far to get to cases of complexity overflow that *do* matter. In fact, the front page of any day's newspaper is quite far enough. Here you'll find headlines telling about the latest installment in the ongoing saga of the teetering global financial system, the failure of safety mechanisms in nuclear power plants, and/or the collapse of trade and tariff talks aimed at restarting the process of globalization. These stories would be enough to give anyone the creeps. Even scarier, though, is that what's been publicly reported is far from an exhaustive list, as the pages of this book will attest.

Complexity science as a recognizable discipline has been with us for at least two decades, so what's the urgency in bringing this message about complexity and extreme events to the public's attention now? The reason is very simple. Never before in the history of humankind has our species been so vulnerable to a gigantic, almost unbelievable, "downsizing" in our way of life than it is today. The infrastructures required to maintain a postindustrial lifestyle—power, water, food, communication, transportation, health care, defense, finance—are so tightly intertwined that when one system sneezes, the others can quickly get pneumonia. This book outlines the dimensions of the problem(s) we face today, how they arise, and what we can possibly do to reduce the risk of a total system meltdown, where in this case human civilization itself is "the system."

THE ORDINARY AND THE SURPRISING

As a filler of time between birth and death, life for an individual, a nation, or even a civilization boils down to a long chain of events. To paraphrase a well-known saying, it's just one damned event after another. Most of these events are inconsequential. You order a steak at the restaurant instead of lobster; that's an event for you and for the chef, who has to prepare the meal. The City of Vienna decides that road traffic will be banned on Graben; that's an event that has long-lasting effects on those of us who live in the First District of Vienna and for tourists, but for almost no one else. The decision of the American government to invade Iraq is an event of major impact on the entire world for decades, maybe longer. Most such events, regardless of the level and magnitude of their impact, are rare in the sense that our expectation of seeing exactly that occurrence—and not something else—prior to its actually taking place, is small, negligible actually. But they are not in any way *extreme*. If it's X-events you're after, then it's the degree to which the event's occurrence is surprising within the context in which it takes place, together with its impact on society as a whole, that matters.

Let's take a moment to deconstruct these two defining aspects of an X-event, starting with the fact that it is by definition an outlier occurrence.

X-EVENTS

WHEN THE WEATHERMAN SAYS THAT THE CHANCE OF RAIN TOMOR-row is 60 percent, he means that the meteorological model he's using predicts that the projected temperature, wind velocities, and the like for tomorrow have in the past led to rainfall 60 percent of the time. So the weatherman is statistically processing the historical record of meteorological quantities, looking for the fraction of the time that rain ultimately came pouring down.

The idea of statistically processing past historical data underlies not only weather forecasting, but a large number of prediction methods and techniques for other phenomena as well. But—and it is a *big* "but"—the historical record must be rich enough and broad enough to encompass the event whose likelihood we're trying to estimate. What if it isn't? What if the historical record is too short, too thinly populated, or simply does not contain something even remotely similar to our target event? What happens then? How can we get a handle on the chances of our specific case turning up? This is the domain where "rare," "improbable," and "unlikely" morphs into "surprising." And the more surprising, the greater the extremity—and potential "X-ness"—of what actually takes place. Here's an excellent example of how to address the issue of the surprise value of an event when the database of possibilities is too small to encompass the behavior in question.

• • •

EVERY SPORT HAS ITS DEFINING MYTHIC ACHIEVEMENTS, A PERFOR-mance that by common consensus will remain in the record books until they crumble to dust. For American baseball, one such legendary event is the fifty-six-game hitting streak set by Joe DiMaggio during the 1941 season. Folk wisdom has it that this streak of fifty-six con-secutive games with at least one hit was essentially impossible. The same goes for its chance of being broken, as it routinely appears on lists of "unbreakable" or "untouchable" records. But the streak *did* happen. So how likely was it, really? Was it the once in the lifetime of the universe fluke that most baseball aficionados believe? Or on a second Earth on the other side of the galaxy could it have happened many times over the last seventy years?

A while back, Cornell researchers Samuel Arbesman and Stephen Strogatz decided to tackle this question. To do so, they envisioned ten thousand parallel Earths, all with the same players having the same statistical performance records—but subject to different whims of chance in each Earth. In essence, what they did was replay each and every one of the seasons from 1871 to 2005 ten thousand times, looking for the longest hitting streak in each of those seasons. Instead of asking how rare DiMaggio's particular hitting streak was, the Cor-nell investigators asked a vastly more general and interesting question: How surprising would it be for anyone in the history of baseball (up to 2005) to have had a hitting streak of at least fifty-six games? Answer: Not very surprising, at all!

In the ten thousand parallel seasons, the longest hitting streaks ran from a modest 39 games to an astonishing (and definitely rare!) 109 games. More than two-thirds of the time the longest hitting streak was between fifty and sixty-four games. In short, there was nothing very extreme about a fifty-six-game streak. In an odd numerical coin-cidence, DiMaggio ranked only as number fifty-six on the list of the most likely player to hold the record for the longest streak in baseball history. Who's the most likely? For baseball fans only, the winners in this derby are two old-timers, Hugh Duffy and Wee Willie Keeler, who between them set the record in more than one thousand of the simulated seasons. For more-or-less modern players, the winner was Ty Cobb, who had the longest hitting streak in nearly three hundred of the ten thousand seasons.

What matters for us in this book is that an event that seems so rare as to be accorded the label "mythical" may in fact be quite humdrum—in another universe from ours! The problem is that our "single-Earth" database may be just too small to be able to say how rare something *really* is. So what is and isn't an X-event is a relative, not an absolute, property of any particular event, and its rarity depends on the context. Just because you and your golfing buddies think it's rare doesn't necessarily make it so.

But even being highly unlikely and surprising isn't quite enough to catapult an event into the category of X-events. For that we need a second ingredient: impact!

It's no exaggeration, I think, to say that memorable events, those that matter, are the ones that in some way change a person's—or a nation's—destiny. This change may be for better or for worse. But game-changer events are, by definition, those that have an impact. Using our weather example, rain tomorrow has little memorable impact for individuals unless they happened to be planning an outdoor wedding or are a farmer worrying about irrigation of his fields. But if the weather turns out to be a tornado, then your life might well be changed as your house is reduced to matchsticks in a minute. In such a case, the surprising event has serious impact—and not for the better. It's fair then to call the tornado an X-event, at least for those affected by it. At a broader level, Hurricane Katrina was both surprising and of huge impact over an area much larger than that of a tornado, and thus is an even bigger X-event. It's not hard to extend this formula of surprise plus impact into the domain of events generated by humans: the 9/11 attacks, the mortgage crisis in 2007–2008, and the 2003 East Coast power failure each may be considered X-events.

• • •

WHY IS IT WHEN THE TERM "EXTREME EVENT" IS MENTIONED WE almost always think of it as characterizing something that is threatening or destructive? Insight into this question comes from looking just a bit harder at three descriptive properties of an X-event.

The common features characterizing all events are an *unfolding time* for the event from its beginning to end, an *impact time* during which the cost or benefit of the event is experienced by some individual or group, and the *total impact* measuring the overall magnitude of the event, usually denominated in dollars or lives lost. (For those readers of an analytical turn of mind, I've included in the book's Notes and References section a single formula allowing us to measure the "X-ness" of an event on a scale ranging from 0 [not an X-event at all] to 1 [the most extreme of all possible events]. I've tried to confine much of the technical details to the notes, but I encourage the strong of heart to explore them.)

When thinking of the term "event," we generally conceive of it as an occurrence of something having a rather short unfolding time, say an auto accident or winning the lottery. This is probably because we tend to have fairly short attention spans ourselves—a feature that's being exacerbated daily as advances in telecommunication and rapid long-distance travel come online. An event occurring quickly (short unfolding time) that generates a big impact having lasting effects (long impact time) is an event that's surprising, difficult to prepare for, and nasty. The March 2011 Japanese earthquake and its attendant tsunami and nuclear power plant meltdown illustrates this sort of X-event. The second law of thermodynamics, which says that unattended systems tend to a state of maximum disorder, tells us that it's always a lot easier and quicker to destroy something than it is to build it. So events with a short unfolding time and large impact, at least at the level of nations and societies, are almost always necessarily destructive.

You might wonder whether there could be "nice" X-events? Yes, there definitely can be! But dilettantes and get-rich-quick schemers take note: such benevolent X-events almost always involve a rather long unfolding time. Think of things like the Marshall Plan that helped West Germany rise from the ashes after World War II or on an even longer timescale, the development of agriculture and the domestication of animals that enabled modern civilizations to develop. Likewise, a breakthrough drug or medical procedure might be the product of years of research, and cultural achievements like a revolutionary novel or artwork are similarly the products of long periods of trial and error. These sorts of scenarios take years, decades, or even centuries to unfold and involve the building up or development of infrastructures, such as a business, a nation, or a technological innovation. So if the examples in the pages to follow seem unrelentingly downbeat, the reader should keep in mind that positive X-events can and do occur—just not in this book! Nice surprises are always welcome. But they are seldom threatening. And it's the threats to our modern way of life we want to focus on here.

Now we have an idea of what constitutes an X-event. Although definitions, even loose ones, are useful, what we really want to know is how such surprises come about and what we can do to either prevent them, or at least prepare for and mitigate their adverse consequences.

SYSTEMS IN COLLISION

In recent years, we've seen long-standing regimes in Tunisia, Libya, and Egypt sent packing almost overnight, with Bahrain, Yemen, and Syria all now being torched by the very same revolutionary flames as rebels battle entrenched governments in an attempt to overturn decades of oppression. On the surface, these types of civil disturbances give the appearance of arising from public discontent with governments over high unemployment, rising food prices, lack of housing, and other such necessities of everyday life. But such explanations are facile, failing to address the root cause of the societal collapse. A civil disturbance is not a game changer by itself but simply a precursor, or early-warning signal, for the coming X-event of regime change. The real source of the X-event that moves regimes sits much deeper in the social system. It is a widening "complexity gap" between a government and its citizens, with revolution breaking out when that gap can no longer be bridged. Think of a rigid authoritarian government confronted with a populace that has awakened to new freedoms through their contact with the outside world and that is coordinated by diverse social-networking platforms. The gap between the complexity of the control system (here the government) and the increased complexity of the controlled system (the population) has to be bridged. One path is for the government to repress the population—imprison leaders, send soldiers to disperse crowds of protesters, and use other measures to tamp down the situation. Alternatively, the government's complexity must be increased so as to speed up the holding of freer elections, quickly remove restrictions on an open media, and open possibilities for upward mobility of the population.

This notion that an X-event is human nature's way of bridging a chasm in complexity between two (or more) systems is the leitmotif running throughout the human-caused X-events I examine in this book. An X-event is the vehicle by which a disparate level of complexity between two (or more) systems in competition or even cooperation is narrowed. In particular, it is the default path when humans themselves fail to voluntarily narrow a widening gap. Let me just give a flavor of how this principle plays out by quickly recalling a couple of recent X-events where this mismatch is especially apparent.

Egypt had a state-controlled economy that was wildly mismanaged for decades. Even the noticeable improvement in the past few years is a case of too little, too late. Moreover, the country was (and still is) monumentally corrupt, as crony capitalism runs rampant throughout the entire social structure. Such a system of corruption relies upon bribes to officials to get contracts, obtain jobs, or to find adequate housing. Amusingly (and tellingly), the impotence drug Viagra was reportedly kept off the market because its manufacturer, Pfizer, failed to pay a large enough bribe to the Egyptian minister of health for its approval.

This type of parasitic mismanagement and corruption worked to freeze in place an already low-complexity government, one that had very few degrees of freedom in either its structure or means to deal with social problems as they arose. But as long as the Egyptian population had even more limited ways to express their dissatisfaction with a lack of proper housing, rising food prices, minimal health care, and the like, the government had no motivation to create the framework(s) necessary to provide these services. Of course, there was a ministry charged with health care, for example. But it served mainly as a sinecure for career bureaucrats and cronies of those in power, and it provided health care only as a kind of spare-time "optional extra." Who would expect this to ever change as long as the spectrum of actions available to the citizenry was kept at a low level (low complexity) much lower than that of the government itself? But times change. And when modern technologies like instant global communication, widespread higher education, and rapid transportation started making their way into the Arab world, citizens quickly became empowered. At this point, the handwriting was on the wall (more to the point, Facebook walls) for entrenched regimes throughout the region.

Modern communication and social-networking services like Google, Twitter, and Facebook serve to dramatically increase social complexity—but now it's the complexity of the population at large that's enhanced, not that of the government. This fact is why such services are routinely restricted or even shut down when governments are under attack, as when the Egyptian government totally closed the Internet for a few days, since these services allow more voices to be heard and more highly connected social networks to be formed. At some point, the gap between the stagnant level of government complexity and the growing level of complexity of the general public becomes too great to be sustained. Result? Regime change in Tunisia, Libya, and Egypt, along with the likely downfall of the Assad dynasty in Syria and/or the monarchy in Bahrain.

A complex system theorist recognizes immediately that the principle at work here is what's called the *law of requisite complexity*. This "law" states that in order to fully regulate/control a system, the complexity of the controller has to be at least as great as the complexity of the system that's being controlled. To put it in even simpler terms, only complexity can destroy complexity. An obvious corollary is that if the gap is too large, you're going to have trouble. And in the world of politics, "trouble" is often spelled "r-e-v-o-l-u-t-i-o-n."

Examples of such mismatches abound. Take the Roman Empire, in which the ruling classes used political and military power to control the lower classes and to conquer neighbors in order to extract tax revenues. Ultimately, the entire resources of the society were being consumed simply to maintain an ever-growing, far-flung empire that had grown too complex to be sustained. The ancient Mayan civilization is another good case in point, as is the former Soviet Union. Some scholars, including historian Paul Kennedy, have argued that the American empire, which spends over $23 billion a year on foreign aid and consumes far more than it exports, is in the process of coming undone for much the same reason.

This type of mismatch is not confined just to complexity gaps in political and governmental domains either, as evidenced by the disruption of everyday life in Japan arising out of the radiation spewing forth from the Fukushima Daiichi reactors damaged by the March 2011 earthquake. The ultimate cause of this social discontent is a "design basis accident," in which the tsunami created by the earthquake overflowed the retaining walls designed to keep seawater out of the reactor. The overflow damaged backup electrical generators intended to supply emergency power for pumping water to cool the reactors' nuclear fuel rods. This is a twofold problem: First, the designers planned the height of the walls for a magnitude 8.3 quake, the largest that Japan had previously experienced, not considering that a quake might someday exceed that level. What's even worse, the generators were located on low ground where any overflow would short them out. And not only this. Some reports claimed that the quake itself actually lowered ground level by two feet, further exacerbating the problem. So everything ultimately hinged on the retaining walls doing their job—which they didn't! This is a case of too little complexity in the control system (the combination of the height of the wall and the generator location) being overwhelmed by too much complexity in the system to be controlled (the magnitude of the earthquake and its consequent tsunami).

Right about now, a conventional risk analyst at an insurance company or a bank might be asking, What's new here? If we want to assess the risk of particular event Y taking place, we calculate the probability that Y happens, evaluate the damage done if Y does indeed occur, and multiply these two numbers together. That calculation tells us the expected damage if Y happens. And that *is* the risk. No muss, no fuss. So what am I suggesting here that differs in any important way from this? For those readers who skipped over the last several pages of this introduction, let me summarize why this question from the conventional risk analyst is the right question to ask—for "normal" events. But it's far less than the right question, a dangerous one even, to ask and assume the answer is nothing when we start talking about extreme events. Here's why.

First, the very rarity of an X-event means that we do not have a database of past actions and behaviors sufficiently rich to actually calculate a *meaningful* probability for the event Y to actually take place. While probability theorists and statisticians have developed an ingenious array of tools ranging from subjective probability theory to Bayesian analysis to extreme-event statistics to try to circumvent this obstacle, the fact remains that nailing down a probability you can believe in for a rare event to occur is just not possible. If it were, we would not have experienced things like the Great Recession of 2007–2008, the 2003 East Coast power failure, and the damage to New Orleans from Hurricane Katrina. And people would not be wondering when the next game-changing shock is going to come jumping out of the closet and kick us in the you-know-where. So when it comes to X-events, we need to invent/discover ways to measure risk that capture what we mean when we say that this shock is more likely to take place here and now than it was previously. My take on this problem is to advance the idea that the magnitude of the complexity mismatch between interacting human infrastructure systems serves as just such a measure.

The second ingredient involved in a conventional risk analysis of normal happenings is the damage the event inflicts on society, should it take place. The only difficulty is that if it's a shock that's never happened before, then assessing actual damage will be highly problematic. To make such an evaluation, we generally need to compare a given hypothetical scenario with similar comparable events from the past. But how could this process work if there is no historical record to draw upon? As shown throughout the course of this book, when the real world doesn't supply the necessary data, we must often build a surrogate world in our computers to obtain that data, much as Arbesman and Strogatz did to study the case of Joe DiMaggio's "unbreakable" hitting streak in baseball. This, again, is an approach very different from what's employed to study normal events.

In summary, there are two very different regimes here. There's the *normal regime,* consisting of events that have taken place many times in the past and for which there is good data available upon which to base our tools for generating probabilities and possible damages. Then there is the *X-events regime* for which those tools simply do not work. This book offers a perspective for creating a framework to systematically study the X-events regime, a framework that complements what's used to calculate risk in the normal regime. I present this line of argument both by precept and by example over the next couple of hundred pages, leaving the technical details for a research program to be carried out in the years to come.

With matters of complexity gaps and their consequent X-events now in hand, let me turn to a very brief overview of the three parts constituting this book.

THREE EASY PIECES

THE OPENING SENTENCE IN JULIUS CAESAR'S *DE BELLO GALLICO* states, "All Gaul is divided into three parts." So too this book. Part I explores the relationship between complexity and X-events, in which the scaffolding I've set up in the preceding pages is developed much further. Here I separate those nasty surprises thrown our way by nature from those generated by human inattention, inaction, misunderstanding, stupidity, or just plain malevolent intent.

The main course, Part II, is served in eleven bite-size chunks, each of which tells the story of a possible X-event and its impact on our everyday lives, should it actually take place. I've chosen these examples to cover as broad a range of human activity as possible, while trying to avoid retracing territory that's already been well chronicled in recent years by the X-event du jour, such as the global financial collapse of 2007–2008 or the 2011 Japanese nuclear reactor crisis. So the elements in Part II range from a collapse of the world food-supply system to a full-scale meltdown of the Internet and from there to a global pandemic and even the disappearance of globalization. For the most part, these stories can be read in any order as the reader's tastes and interests dictate. Taken together, though, they paint a picture of just how varied and serious is the threat of an X-event to the sustainability of our accustomed style of life.

The finale, Part III, brings the theoretical questions and problems of Part I together with the graphic examples of Part II, in order to come to grips with the central question of how to anticipate X-events and even manage them—sometimes. More specifically, I examine the degree to which we can sharpen the focus in both time and space for when a potential game changer of a certain type is entering into the danger zone of actual realization. I also investigate the types of "weak" signals that serve as tip-offs to such an impending shock, along with methods for ferreting out these signals from the avalanche of noise that passes for information on a daily basis. The book concludes with some precautionary advice as to how societies might better prepare for X-events through the creation of more resilient social systems and less fragile infrastructures.

PART I

WHY NORMAL ISN'T SO "NORMAL" ANYMORE

OUT OF NOWHERE

THE BIG NEWS STORY IN AUGUST 2011 WAS THE THREAT HURRICANE Irene posed to the way of life of New Yorkers. If events happened to fall just "right," Manhattan and other low-lying regions might well be swamped, shutting down public transit systems, the New York Stock Exchange, food and water supply chains, and other niceties of urban life in New York City. Luckily, Irene ran out of steam before delivering such a life-changing blow, and the end result was a just bit worse than what would have been experienced from a very heavy summer storm. So the good people of Manhattan dodged a bullet, and the media hype beforehand simply amounted to a large dose of overreaction to the possible threats presented by Irene. But one day the bill really will come due, as it did in New Orleans with Hurricane Katrina in 2005, showing that overreaction and near-paranoiac prudence are two sides of the same coin.

Of course, believing that the sky is falling is a well-known defense mechanism when we're faced with a threat so far outside our comfort and knowledge zones that we can only run around like Chicken Little, hoping the threat goes away. Often it does. But not always. And it's those "not always" cases that don't just capture headlines, but also force us to confront the unpleasantries of existential threats to an accustomed way of life. This is no joke, either. X-events do occur. And they can wreak the kind of havoc and damage that many of us would like to believe only science-fiction writers and Hollywood film producers take seriously. But an X-event is not reality television; it's reality, period. Here are a couple of examples as prelude for the stories told in the pages of this book.

ABOUT 74,000 YEARS AGO, ON WHAT IS NOW THE ISLAND OF SUMATRA in Indonesia, the supervolcano Toba exploded with a force that cannot be compared with anything that has been seen on Earth since the time humans began to walk upright. Just for the sake of comparison, the eruption of Krakatoa in 1883 had an explosive force of 150 megatons of TNT, which was ten thousand times greater than the atomic bomb that flattened Hiroshima. The Toba explosion is estimated to have been about one *gigaton*, over six times greater than Krakatoa and three thousand times greater than the energy produced by Mount St. Helens when it erupted in 1980.

At the time of the Toba event, Neanderthal man inhabited Earth, alongside *Homo sapiens* in Europe and *Homo erectus* and *Homo floresiensis* in Asia. The last Ice Age was then at its peak, with woolly mammoth and saber-toothed tiger on the menu for humans of the day. The volcano changed all that—overnight.

Alongside gigantic tsunami waves, the nearly three thousand cubic kilometers of dust and debris ejected into the atmosphere reduced the solar radiation so dramatically that plants received too little light to survive. The average temperature worldwide dropped to (note: not *by*, but *to*) 5 degrees (Fahrenheit), turning summer to winter and winter into a very deep freeze.

Today it's estimated that the total number of human survivors of this event was only a few thousand, mainly those living in small groups in Africa. This figure is the result of remarkable genetic detective work by scholars who examined DNA samples from that period. The researchers saw that the genetic samples from all over the world would have been very different if humans had been able to develop without the difficulties Toba created everywhere on the planet. Today's humans all stem from those hardy survivors. It was science journalist Ann Gibbons who suggested in 1993 that the Toba volcano was responsible for the near extinction of humankind, a hypothesis that was quickly taken up by researchers including Stanley Ambrose of the University of Illinois, who developed theories and dug up the data to back up Gibbons's idea.

But still "almost" only counts in horseshoes, and even a supervolcano like Toba couldn't in fact totally wipe humans from the face of the planet. It was a monumental catastrophe. No doubt about that. But it didn't send humankind to the graveyard of history. What *could* do that?

To get a handle on what type of event might end the tenure of *Homo sapiens* on Earth, we need go no farther than to the most popular wing in your local natural history museum. About sixty-five million years ago, a fireball six miles in diameter crashed into what today is the Yucatan Peninsula in Mexico at a speed of about twenty miles per *second*. This "near-earth object" (NEO) created some of the same existential threats to life a supervolcano generates—fireballs, tidal waves, blast effects, and the like—but on a scale that dwarfs even the biggest supervolcano. To get a sense of what it would be like if such a behemoth struck a landmass today, here's a plausible scenario.

First, the local explosion would literally annihilate everything in the immediate area of the impact, generating a shock wave radiating outward for hundreds of miles that kills everything in its path. Gigantic fires would spread for as much as five hundred miles in all directions. And this is not to speak of the global firestorm produced by the huge amount of debris that would be sent into the atmosphere and come raining down all over the planet. The energy released from the strike would probably raise surface temperatures on Earth to levels seen inside a hot oven. It's important to note that debris in the atmosphere would block the sun's rays, producing a global freeze sometime after the impact and that massive shock waves traveling through the planet from the impact could well trigger volcanic activity as an aftereffect.

These effects are more than enough to have led to the demise of the dinosaurs, who dominated all life-forms on Earth for a whopping 170 million years. Their disappearance opened an eco-niche for some small, furry mammals about the size of a big rat that ultimately evolved into . . . today's humans. The one thing that's for certain, though, is that an object like this striking the planet today would stop civilization in its tracks. It's more than conceivable that every land animal larger than a cat would die. But who knows? After all, no dinosaur could lay in a supply of canned food, sacks of corn, or ensure a supply of potable water in a deep underground shelter. So some people might conceivably make it through even an event of this magnitude. But it doesn't seem to be the way to bet. Besides, who wants to live in a world in which the survivors would almost surely envy the dead?

In our current form, we humans have been in place for at most a few hundred thousand years. Of course, an asteroid like the one that struck the Yucatan itself occurs only every few hundred thousand years or more. But, then, what's a few hundred thousand years compared to the nearly two hundred million enjoyed by the dinosaurs before they exited center stage?

So there we have it: the reality of disaster, catastrophe, extinction. Take your pick. It will not have escaped the perceptive reader that each of the X-events I've so far outlined share a common cause: nature. Earthquakes, volcanoes, asteroid strikes, and other such events are all outside the scope of either human cause or human intervention. We are pretty much powerless to influence whatever nature wants to throw at us. And if we get very unlucky, we'll hang up a GOING OUT OF BUSINESS sign, turn out the lights, and simply declare that the party's over. So while these sorts of interventions by nature are useful as background to the story I tell in this book, by far the more interesting and relevant part of this story for humankind is the other side of the X-events coin: human-caused catastrophes, perhaps aided and abetted by nature. Let's look at a few examples of human-caused X-events that are analogous, but vastly smaller in impact, than the examples we've just explored. I begin with some hypothetical X-events illustrating the spectrum of possibilities, then move to examples of the real thing.

THE HUMAN FACTOR

CONSIDER THE FOLLOWING POSSIBLE X-EVENTS:

- A virulent strain of the avian virus jumps to humans in Hong Kong, sweeps across Asia, and ends up killing more than fifty million people.
- A magnitude 8 earthquake centered on the Ginza in Tokyo kills two million people outright and leads to property damage running into the trillions of dollars.
- Bees around the world begin dying off in massive numbers, interfering with pollination of plants worldwide, thereby setting off a global food shortage.
- Terrorists detonate a nuclear weapon in Times Square during rush hour, leveling much of Manhattan, killing half a million people, and permanently reducing New York City to rubble.
- A tanker car carrying chlorine derails in Rio de Janeiro, spilling its contents and killing more than five million Cariocas.

This list could be carried on almost indefinitely. The point is that surprising events capable of killing millions, if not hundreds of millions, of humans are well within the realm of the possible. Moreover, even without huge loss of lives, capital stock is decimated, setting back development worldwide for decades. Not a single one of the items on this list is impossible. And, in fact, some of them, like the spill of a deadly chemical, have already happened many times.

Today humans are more vulnerable than ever to X-events. The complex infrastructures we depend upon for everyday life—transportation, communication, food and water supply, electrical power, health care, to name but a few—are fragile beyond belief, as we're reminded when even a small glitch in the delivery systems occurs. What are the causes of this high fragility and our consequent vulnerability? Can we really understand these X-events and if not actually control them, at least anticipate them? To address such questions, we need to have some understanding of the root cause(s) that give rise to these events and whether these causes are hardwired into the workings of the infra-structures themselves or are something we can forecast and to some degree control.

As I argue throughout the remainder of the book, the under-lying cause of X-events is directly attributable to the ever-increasing complexity of our global society. This complexity shows up in many forms, such as the high connectivity among infrastructures that trans-mits a tremor in one part of the infrastructure to other parts of the system, often at literally the speed of light. Sometimes the complex-ity reveals itself in layers of bureaucracy piled upon existing layers until the system can no longer function—what I'll term "complexity overload" in the pages to follow. At other times, the problem mani-fests itself not with any single infrastructure, but as a "mismatch" in complexity levels between two or more interacting systems, such as a nation's government and its citizens, as discussed earlier. But in all cases, the systems we count on for everyday life cannot function if they are beyond the ability of their regulator to understand. When the complexity level or the mismatch becomes greater than the system can bear, a reduction is needed to rectify that situation. An X-event is simply the system's way of restoring a sustainable balance.

This balancing act will be our leitmotif. Our fortunes are tied to it. If it fails, humanity fails. What's really worrying is that the sys-tems supporting our twenty-first-century lifestyle are simply not as robust as we'd like to think. On this note, let me present a handful of slightly more extended examples of human systems breaking down in ways that hint at how vulnerable we are should a complexity overload highly stress the system.

• • •

ACCORDING TO A 2004 REPORT IN THE *LOS ANGELES TIMES*, A MAJOR breakdown in Southern California's air traffic control system was due partly to a "design anomaly" in the way Microsoft Windows servers were integrated into the system. The radio system shutdown, which lasted more than three hours, left eight hundred planes in the air without contact to air traffic control and, according to the Federal Aviation Administration, led to at least five cases where planes came too close to each other. Air traffic controllers were reduced to using their personal mobile phones to pass on warnings to controllers at other facilities, and they observed numerous close calls without being able to alert the pilots.

The FAA ultimately concluded that the system crash was attributable to a combination of human error and a design problem in the Windows servers brought in over the past three years to replace the traffic control system's original Unix servers. The servers are timed to shut down after 49.7 days of use in order to prevent a data overload, a union official told the *Los Angeles Times*. To avoid this automatic shutdown, technicians are required to restart the system manually every thirty days. An improperly trained employee forgot to reset the system, leading it to shut down without warning, the official said. Backup systems failed because of a software glitch.

Three years later in June 2007, a computer system in Atlanta that processes pilot flight plans and sends them to air traffic controllers also crashed, setting off a cascade of such failures around the country. These breakdowns led to hundreds of flights being delayed and even canceled altogether in both directions from New York airports. Just one year later, the very same computer in Atlanta froze again. The problem occurred as routine software work was being done on the computer, leading it to offload work to the other control system located in Salt Lake City. But the Utah system was overwhelmed by the surge of information and could no longer process all the flight plans being filed. End result? No flight plans were processed, which meant no controllers knew what routes planes would be taking and when they intended to land. At that point, controllers stopped giving takeoff clearance and the whole air traffic system froze up.

But human error in the skies is certainly not confined to misunderstandings and aging computers. In September 2010, a US Airways jet carrying 95 people came within fifty feet of a small cargo plane while taking off from Minneapolis, and just a couple of months later an American Airlines flight with 259 aboard nearly collided southeast of New York City with two air force transport planes. Later, an air traffic controller at the Ronkonkoma, New York, radar center that was controlling the American plane complained about the unprofessional, sloppy atmosphere at the center. And this is no backwater facility either, but the second-busiest air traffic radar center in the country.

These stories could be multiplied severalfold, as well as spiced up, by accounts of controllers falling asleep in the tower and other purely human frailties, all of which add up to increasingly less friendly skies for today's air traveler. The statistical data bear out this scary picture, as reports of mistakes by air traffic controllers nearly doubled from 2009 to 2010 with no end in sight. Fortunately, most of the errors were not in the most serious category, the sort that would require pilots to take evasive action. But even in that class, the errors reported rose from thirty-seven in 2009 to forty-four in 2010. All in all, there still remain good reasons to be seriously concerned about a true X-event occurring that might shut down the entire air traffic system. The system is highly fragile, perched on the edge of a complexity gap between the airlines wishing to maximize flights and cluster them during desirable travel times and the need by controllers to keep the skies safe. As these stories suggest, the gap seems to be widening at an alarming rate.

• • •

ON FEBRUARY 24, 2010, GREEK POLICE FIRED TEAR GAS AND CLASHED with demonstrators in central Athens after a march organized by unions opposing the government's program to cut the European Union's biggest budget deficit. The president of a large union stated, "People on the street will send a strong message to the government, but mainly to the European Union, the markets and our partners in Europe, that people and their needs must be above the demands of markets. We didn't create the crisis." Later, air traffic controllers, customs and tax officials, train drivers, doctors at state-run hospitals, and schoolteachers walked out also to protest government spending cuts. Journalists joined in to the strikes as well, creating a media blackout that further exacerbated the situation.

This brouhaha in Greece was a prime example of a civil unrest turning into a civil disturbance, calling for intervention by authorities to maintain public safety. For the sake of clarity, a civil disturbance might take many forms: labor unrest, strikes, riots, demonstrations, up to actual rebellions leading to political revolution. Events triggering such disturbances range from racial tension and religious conflicts to unemployment, unavailability of goods and services like food, water, and transportation, or unpopular political actions like launching the wars in Vietnam and Iraq.

A cursory examination of how an increasingly negative mood in a population can lead to social unrest suggests many ways that civil disturbances might easily emerge. The Greek situation above is an excellent illustration of what can happen when a relatively minor event in the financial sector occurs. If that incident takes place at a moment when a country is poised on the brink of disruption like Greece (and maybe Spain, Portugal, or Italy, as well), a small shove (in this case from the European Union financial authorities) can push it over the edge. The events of January 2011 in both Tunisia and Egypt, and later in Libya and Syria, simply underscore this point.

One might wonder what a truly major shock, as opposed to a gradually deteriorating social mood, might do. For instance, consider the

January 2010 earthquake that destroyed Port-au-Prince, the capital of Haiti, a country with one of the lowest per capita income levels in the world. Or imagine the social disruption that would be created by an Internet blackout lasting several days in a major population center in the industrialized world like London, Tokyo, or New York. This would shut down electrical power, transportation systems, food supplies, and communications, not to mention having a disastrous impact on banking and web-based businesses. The almost inevitable looting, rioting, and other forms of social unrest to follow would make what happened in Athens look like polite teatime chatter at the Queen's garden party. Or suppose a Black Plague–style pandemic were to break out in a densely populated city like Hong Kong or São Paulo. Again, the mind boggles at the havoc that would ensue. Currently, what seems to be the most likely candidate to create this sort of disruption is another financial tsunami.

In a report on worldwide economic prospects for 2010, the credit-rating agency Moody's warned that countries with fast-rising government debt must steel themselves for a year in which "social and political cohesiveness is tested." The report raised the prospect that future tax increases and spending cuts could trigger social unrest in a range of countries in both the developed and developing world. As an advance look at the possibility of a financial crisis in a leading economy, the report went on to say that 2010 would be a "tumultuous year for sovereign debt issuers." Looking at what actually unfolded in 2010 and 2011, this statement appears to be about as prophetic as it gets. Extrapolating just a bit, what might we expect in the years to come?

A good guess is that as people lose confidence in the ability of their governments to solve the financial crises, they'll break out into protests and/or assaults on those they see as responsible for their misery. This group will certainly encompass government officials and bankers, but it may well include immigrants, ethnic and religious minorities, landlords, and even corporate managers and bosses as well. The Occupy Wall Street movement in late 2011 is an excellent illustration of this process in action. If you want to be grimly impressed, start putting pins on a map where such violence has already broken out; cities like Athens, Sofia (Bulgaria), Port-au-Prince, Riga (Latvia), and Vilnius (Lithuania) are immediately marked, as do Tunis, Cairo, Damascus, and Sana'a. Even much larger cities like New York (the "Occupy Wall Street" demonstrations), Moscow, Rome, London, Paris, and Dublin have seen huge protests against rising unemployment and declining wages, as well as outrage over the yawning gap between the rich and "the other 99%." But security police in these cities have managed to keep the protests orderly, if not peaceful (so far).

One might even characterize what's going on here as a global "outbreak" of economically driven violence, a kind of societal "pandemic." While it seems likely that these disruptions will be confined to specific locales, we cannot entirely discount the possibility that as the global economic situation worsens, some of these localized incidents will overrun national borders and lead to far more widespread and long-lasting events. Armed rebellions, military coups, and even wars between states over access to resources cannot be ruled out.

But even social disruption of the revolutionary variety pales in comparison to what a combination of nature and humans together can brew up. So let me conclude this highly abridged catalog of X-events with two more examples, each a prelude to a more extended account of the X-event presented here in two of the chapters to follow in Part II.

ONE OF THE BEST-SELLING BOOKS OF 1969 WAS MICHAEL CRICHTON'S novel *The Andromeda Strain,* which tells the tale of a team of scientists studying an extraterrestrial microorganism that quickly—and fatally—clots human blood. This was Crichton's breakout book, crowning him king of the techno-thriller writers. While patently fiction, *The Andromeda Strain* gives a chilling account of the biological threat posed by organisms that the human immune system has had no exposure to and thus cannot combat. In the book, the organisms come from outer space; in real life, they may come from "inner space," via either accidental or deliberate human biotech activity.

To illustrate the possibilities, some years ago a group of Australian researchers produced a strain of mousepox, a variant of the smallpox virus, hoping to make the mice infertile. Normally, mousepox is not dangerous to the type of mice involved in the experiment, and the scientists just wanted to give it a bit more "juice" to have it act so as to sterilize the mice. Unfortunately, they ended up with a strain of the virus that was so deadly that it even killed mice that had been vaccinated against mousepox.

This is a prime example of how a laboratory miscalculation can lead to a strain of virus like smallpox that could easily start an uncontrollable pandemic should it get out of the lab and into the wild. And this is no Michael Crichton–like fiction, either, especially when researchers like the Australians publish the formula for their killer virus in the open scientific literature for all to read and possibly replicate in their own labs.

Of course, one can argue that smallpox has occurred before and it didn't wipe humanity from the face of the earth. But those were isolated incidents, not the result of a committed, organized effort to spread the disease far and wide.

A small, but informative, real-world example of what could happen on a broader scale is the Spanish flu epidemic that followed World War I. In 1918, a strain of influenza appeared in the United States that ended up killing anywhere from thirty million to fifty million people worldwide within a year. Now imagine a plague or virus that sweeps the world like the epidemic of 1918, but spreads faster and kills those infected more rapidly. There is no vaccine or antibiotic to combat it. Let's also note that this is something that could arise by natural processes, not via directed mutation in a lab. So the threat of a global pandemic moves up from a mere disaster to the level of a true catastrophe.

Nevertheless, it's very likely that just like in *The Andromeda Strain*, where a Sterno-drinking derelict and a crying baby both survived the otherworldly organism, some humans in isolated communities or with a freakish natural immunity will survive just about anything that humans or nature throws at them. So humanity will almost surely survive even the deadliest pandemic. To get to true extinction from a human-caused source we need to ratchet things up a bit.

IRVING LANGMUIR WAS A NOBEL PRIZE–WINNING PHYSICIST WHO worked at the General Electric Research Center for over forty years, retiring in 1950. In 1953, Kurt Vonnegut published the now-classic science-fiction novel *Cat's Cradle,* in which he based the protagonist Dr. Felix Hoenikker on Langmuir, a man Vonnegut knew through his job as a publicist at GE before he became a full-time writer. Vonnegut once said to a journalist, "Langmuir was absolutely indifferent to the uses that might be made of the truths he dug out of the rock and handed out to whomever was around."

In *Cat's Cradle,* Hoenikker creates a substance called "ice-nine," an alternate form of water that is solid at room temperature. When a crystal of this strange substance is brought into contact with normal liquid water, it serves as a template through which the molecules of normal water rearrange themselves into solid form. The story portrays Hoenikker as basically amoral and completely uninterested in anything but his research, ice-nine being nothing more than a mental puzzle for him. Ultimately, a Caribbean dictator obtains a crystal or two of this stuff and uses it to commit suicide by instantly freezing his body into a block of solid ice at room temperature. An airplane then crashes into the dictator's palace, causing the still-frozen body to fall into the sea. This, in turn, sets off a gigantic chain reaction in which all the earth's liquids (including blood) turn into ice-nine, leading to the extinction of every living object.

Pretty amazing stuff, ice-nine. But the scenario painted by Vonnegut is not much removed from what some scientists see arising as a live—though still very much an outlier—possibility through work going on today in the rapidly growing area of nanotechnology.

There is no logical or physical reason why self-replicating nano-robots, just a few atoms in size, couldn't become a cancer to the earth's biosphere and very quickly replace all carbon-based life-forms with nanotech life. The primary obstacle to such a runaway cancer is the availability of the necessary energy, since there's plenty of organic matter around for the "nanobots" to cannibalize. It seems as if the nanobot would have to use sunlight or possibly organic tissue to continue its ice-nine-like romp through the ecosphere. But calculations done by numerous nanotech researchers show that the nanocancer should be able to capture and utilize at least half the planet's incident solar energy so as to cover the earth with a kind of "gray goo." So it appears that the only real way to stop this process would be to cut off the energy source and/or interfere somehow with its replication mechanism.

Here I'm not talking about some kind of disaster localized in space and time, like an earthquake or even a catastrophe that wipes out hundreds of millions of people around the world, like a global pandemic. The gray goo, just like Vonnegut's ice-nine, simply destroys the entire ecosystem that all living things today depend upon for their existence. The nanocancer, then, is literally an existential threat to *all* life as we know it today, a full-fledged extinction event generator.

Up to now, I've used the words *disaster, catastrophe,* and *extinction* rather loosely to characterize the various examples we've considered. Before going into this in more detail, it's useful to step back for a page or two and look at the issue of what constitutes an X-event.

WHITE DOVES AND BLACK SWANS

OUTLIER EVENTS LIKE A KILLER DISEASE, A HURRICANE, OR A FINAN-cial crisis seem to happen every week. So how can we call them "outliers" or "extreme" events? And if they happen so frequently, why do we think of them as being "rare"? And why can't we do a better job of forecasting or, at least, anticipating them? We'll see the answers to these and many other puzzles unfold over the coming pages. The short answer, though, is that if we focus attention on any *specific* domain, for example, extreme weather like a hurricane, then the occurrence of that hurricane is indeed rare in the context of weather events. Now let's broaden our horizons and consider many domains like weather, earthquakes, financial market crashes, pandemics, volcanoes, and other areas in which an extreme event of some kind might take place. Now ask the question: What is the likelihood of an X-event occurring in at least one of these domains tomorrow? It won't surprise you to know that the likelihood is actually fairly high. So *some* type of X-event is indeed happening somewhere just about every day.

Of course, this line of reasoning only shifts the uncertainty about timing of an event in a fixed domain to uncertainty about both the location and the domain where the next outlier will turn up. So there's no free lunch. One way or the other, we're going to have to face the fact that some types of events live in the normal regime and their likelihood can be calculated from past data, while other sorts of events, the extreme ones, live in the X-events' regime and are almost impossible to forecast. The problem is that X-events are the game changers of human life. That has never been more true than it is today, when for the first time we humans have the capacity to actually *create* an X-event that would bring about our own destruction. Since nature is no longer the only grim reaper in the game, we have no choice but to use our technology and tools for analyzing the systems of everyday life to uncover some of the secrets of extreme uncertainty, in order to at least postpone, and possibly avoid altogether, the same fate suffered by the dinosaurs and other bygone species.

With that thought as our call to arms, off we go now into the land in which almost nothing is actually known but where much of our future lives will be determined.

● ● ●

• • •

FOLLOWING HURRICANE KATRINA'S DEVASTATION OF NEW ORLEANS in 2005 from the collapse of the levees protecting the city, General Carl Strock of the US Army Corps of Engineers stated:

> When the project was designed . . . we figured we had a 200- or 300-year level of protection. That means that the event we were protecting from might be exceeded every 200 to 300 years. That is a 0.5 percent likelihood. So we had an assurance that 99.5 percent of this would be okay. We, unfortunately, have had that 0.5 percent activity here.

Strock's claim rests on the assumption that hurricanes the size of Katrina occur with a frequency that can be described by the classical bell-shaped curve, the so-called normal probability distribution. Sad to say for New Orleans (and General Strock), hydrologists and statisticians have known for more than a century that the extreme events falling near the ends of a statistical distribution usually cannot be usefully described in this way. As we all saw with painful clarity in the meltdown of the global financial system in 2008, the normal distribution dramatically underestimates the likelihood of outlier events. The bell-shaped curve works well to forecast the behavior of systems whose output is the sum total of a large number of small-scale events, each of which has no influence on any of the others—that is, they are "independent." To illustrate this point, consider all adult males in the United States and ask, What is the average height of this group and how far away from that average is someone who is five feet, four inches tall? To a good approximation, the height of each male does not depend on the height of any of the others, and the number of males is very large, just the conditions under which the bell curve works well to answer questions like these.

But if we could actually determine the type of probability curve followed by extreme events (which we cannot), that curve would be

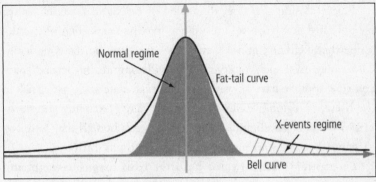

The bell-shaped vs. fat-tailed distributions

one that is informally termed a "fat-tailed" distribution. The difference is shown in Figure 1 below. The normal regime describing many independent events is represented by the traditional bell-shaped curve, the gray line, which seriously underestimates the likelihood of outlier "shocks" in the striped X-events regime. The big outliers live out in these tails. Using this fat-tail law to describe the New Orleans situation, for instance, the 0.5 percent mentioned by General Strock would have been closer to 5 percent and the three hundred years would have shrunk to about sixty years.

Looking at the figure (the only one in this book, I promise), you might think that the striped region of the fat-tail distribution for X-events doesn't really differ so much from that of the bell curve. But this is not the case at all.

To illustrate the implications of the figure, according to the bell-shaped curve the likelihood of a very-large-deviation event (a major outlier) located in the striped region appears to be very unlikely, essentially zero. The same event, though, is several *thousand* times more likely if it comes from a set of events obeying a fat-tailed distribution instead of the bell-shaped one. This means that if an insurance company was selling insurance against unlikely events and based its risk premium on the bell-shaped curve, it might charge a fee of a million dollars. But if the family of events actually followed the fat-tailed probability law, the firm's true exposure could well be several *billion* dollars!

Before closing this brief discussion of bells and fat tails, let me reemphasize that when we use a curve like the one in Figure 1 to describe the likelihood of an X-event, we are speaking metaphorically. There may exist such a curve in some Platonic realm beyond space and time. But we have no way to actually calculate it. To assess risk in the X-events region, we must replace everyday probability and statistics with a new paradigm. My argument in this book is that complexity gaps serve as a starting point for this paradigm shift.

Forecasting models (which form the basis for insurance premiums, building codes, economic expectations, and the like) ordinarily are based only on past data, which is generally a tiny sample of the total range of possible outcomes. The problem is that those "experts" who develop the models often come to believe they have mapped the entire space of possible system behaviors, which could not be further from the truth. Worse yet, when outliers do crop up, they are often discounted as "once in a century" events and are all but ignored in planning for the future. Just as frequently, it's mistakenly believed we have "solved" the problem(s) that a given X-event exposed (think of the waves of legislation that follow disasters) and thus don't have to worry about the possibility of these sorts of outliers anymore. All bases are now covered, so to speak. Unfortunately, there are plenty more X-events where the first one came from. The real lesson here, of course, is that the world is much more unpredictable than we'd like to believe.

Where do these fat tails come from? Let's answer this by looking at the domain where they've been most studied, if not fully appreciated and applied: the stock market. (Note that a "fat tail" is not to be confused with the "long tail" concept in business, in which high inventory and easily accessible vendors like Amazon or iTunes extend the life of "back list" products.)

The key reason fat tails—X-events—exist in financial market returns is that the decisions by investors are not fully independent (this is *the* key assumption that *would* lead to a bell-shaped distribution of market price changes). At extreme lows, investors are gripped with fear and they become more risk-averse, while at extreme market highs, investors become "irrationally exuberant." This type of interdependence leads investors to herd or cluster together, which in turn causes them to buy at ridiculous highs and sell at illogical lows. This behavior, coupled with random shocks from the outside world (possibly X-events in their own right), push market price changes to extremes much more frequently than models based on the bell-shaped distribution would have us believe.

A graphic illustration of this point is that the technical *causa causarum* of the ongoing global financial crisis is the almost universal use of the so-called Black-Scholes formula for pricing asset values for options and other derivative securities. This rule, for which Myron Scholes and Robert Merton received the 1997 Nobel Prize in Economics (Fischer Black having died in 1995), is just plain wrong. Why is it wrong? Again, one of the principal reasons is that it's based on the assumption that investors' decisions are independent, which led the developers to base their mathematics on the bell-shaped distribution. This causes the formula to vastly underestimate the likelihood of high-risk events of the very type that actually occurred in 2007, and which set off the chain reaction of bank failures and financial havoc that plagues us to this day. As one of my friends puts it every year when the Nobel Prize in Economics is announced, "Yet one more reason why there shouldn't be a Nobel Prize in Economics!" If one wanted to nail down the precise moment when this entire line of bell-shaped thinking was exposed as an emperor with no clothes, it's hard to do better than point to the publication of Nassim Nicholas Taleb's best-seller *The Black Swan,* which argued forcefully and provocatively that the whole edifice of theoretical finance rests on quicksand.

IN LATE MARCH 2007, I WAS SPEAKING AT A FORESIGHT WORKSHOP IN Zurich at which Taleb was one of the other invited speakers. His book was just about to be released, and as it happened he received a couple of prepublication copies from his publisher during the course of the meeting and was kind enough to inscribe one of them to me as a sort of memento of our first face-to-face meeting. That occasion offered the opportunity for me to speak with him about the very issue that just a few months later exploded onto the front pages of newspapers around the world: the near collapse of the US financial system, precipitated by the failure of Lehman Brothers in late 2007, and which has since been repeatedly exacerbated by the fumbled-fingered gyrations of the US Federal Reserve and other government political and financial agencies around the world.

Taleb had spent many years as a trader in exotic financial instruments before transiting to the more contemplative life of an academic, public intellectual, and general gadfly. So his perceptive and caustic views on the huge risks being taken every day in financial centers around the world were both enlightening and, I have to say, a bit enervating as well. Here's why.

In his writings, Taleb has described "black swans" as events that "Lie outside the realm of regular expectations, carry an extreme impact, and human nature makes us concoct explanations for their occurrence after the fact." While I don't quibble with Taleb's definition of a black swan—or in our terminology here, an X-event—I think it's incomplete in a couple of meaningful ways. So let me deconstruct his definition to put it into a form somewhat more useful for our discussions here.

First of all, *rarity*. This is certainly the least controversial aspect of what does or doesn't constitute a X-event, as I've already considered in the preamble and revisited in the foregoing pages. X-events are "by definition" outside the realm of everyday expectations. Whether they are totally unexpected, as Taleb's definition suggests, is another matter. After all, even events as rare as an asteroid impact or a 9/11-style terrorist attack are to be expected. The only surprise is when they will happen, where they will occur, and how damaging they will be. But happen they will, regardless of the wishes and hopes and fears of us puny humans to change that state of affairs. Like hurricanes and earthquakes, we can only try to anticipate these game changers and prepare for them so as to mitigate their damage. So we all agree on rarity—but rarity in the sense of being surprising, as discussed in the preamble. Not mere rarity for its own sake as something "infrequent." I'll take up this point again later in the book.

Things become rather more interesting when we turn to the *extreme impact* aspect of Taleb's trinity, since rarity and impact are properly separate matters. A Force 5 hurricane that drowns New Orleans is one thing; the same hurricane harmlessly blowing itself out over the Caribbean is something else again. To a hurricane specialist, the two events are equally interesting; to CNN, (re)insurance companies, and of course the residents of New Orleans, the two cases are very different matters indeed. The dissimilarity, of course, resides in the impact of the event, be it measured in dollars, lives lost, and/or psychic stress. So both rarity and impact have to go into any meaningful characterization of how black any particular swan happens to be.

By far, the most interesting component of Taleb's trichotomy is the one addressing the after-the-fact *stories* we humans tell ourselves by way of explaining an X-event. Of course, this leg of the triangle is clearly the part Taleb relishes most, since humanity's predilection for self-deceiving narratives gives rise to our illusion of being able to both forecast and even control the black swans. In accord with Taleb, I do not believe that there is any person or method, living, dead, or yet to be born, that can reliably and consistently forecast *specific* X-events. By "forecast" I mean predict the timing and location of an event with enough accuracy to be useful in preventing, preparing for, and/or surviving the predicted event. Such a forecast might be something like, "An earthquake of magnitude 6.7 centered near Chula Vista will hit Southern California on February 24, 2017, at 7:47 P.M." To believe such a prediction is possible, even in principle, is to succumb to a hubris that's both dangerous and totally ill-conceived. Forecasts of this type occasionally work in the natural sciences, primarily in astronomy and engineering, and without exception involve events in the regular regime of Figure 1, events that generally occur over a short period of time in a limited geographic area. Thus, I concur with those who say that attempting to truly forecast an X-event is a fool's errand.

On the other hand, I believe it's perfectly feasible to develop tools for *anticipating* X-events, turning a black swan into something that's more like a common white dove. (If I didn't, I wouldn't have written this book!) To understand what this means, bear in mind that X-events, especially those that are human caused, unfold as a combination of what the French biologist Jacques Monod termed "chance and necessity." At any given time, there is a societal background, a kind of playing field, within which human actions and behaviors take place. And that terrain is continually shifting, giving rise at a particular time and place to a sociopsychological climate that tends to favor the occurrence of some types of things happening and work against others. One might think metaphorically of this space as biasing the "flow" of events. Sometimes the flow is moving to favor a particular sort of occurrence taking place; sometimes the flow shifts, and that very same outcome becomes far less likely. It does not *require* any particular action to take place, but only biases the likelihood of what may or may not actually occur out of the vast space of possibilities.

The other half of the story is the "chance" component. In a given environment, many things might happen. What actually does take place is determined by essentially random factors (i.e., those without any discernible pattern) at a particular time, factors that force one of the potential events to actually occur. Think of a space of possibilities that looks like a sharp mountain peak surrounded by many valleys. Currently, you're sitting on top of the peak. A random shove pushes you down a slope in one direction, taking you into one of the many possible valleys. Suddenly, all the valleys except the one you're in cease being candidate destinations. But if before that random shove the landscape had shifted in such a way that what had previously been a peak was now a plateau, you would only have been pushed over a bit on the plateau and not been sent into any of the valleys at all. In other words, nothing would have happened. Or the peak may have become asymmetric, in which case it would have taken a bigger shove in one direction to get into a certain valley than to any of the others.

The point here is that what we actually observe is always a combination of the background circumstances of the moment, together with a random element that cannot be foreseen at all. My belief is that there are many different ways to gain insight into the shape of the playing field and its continual undulations, thus obtaining useful information about what type of event, or otherwise, is more or less likely to be seen at a particular moment in time. So any talk about precision "forecasting" of X-events in this book means simply foretelling how the playing field is changing; it definitely does not refer to the prediction of specific events. For that, you need a card layer or a crystal-ball-gazing fortune teller, not a complexity scientist.

Due to the problem of predicting outlier events, they are not usually factored into the design of systems. This makes them especially dangerous because, as we shall see momentarily, the world has become unimaginably complex and our lives in turn have become completely reliant on increasingly complex systems—the very systems, I remind you, which cannot account for outlier events. Let's look at yet a few more examples to hammer home this point.

MANY READERS OF THESE PAGES UNDOUBTEDLY HAVE A HIGH-TECH coffeemaker in their kitchen or office that produces the perfect cup of espresso simply by pushing a button. The beans get ground, tamped, and prewet, after which the boiling water is forced through the grounds at high pressure to give that shot of caffeine we all seem to need to get our motor running in the morning. In short, this machine is a coffee-making robot. You put in the beans, connect it to a water supply, plug it in, and away it goes at the push of a button. But all the automation built into the coffeemaker comes at a price: a huge increase in the complexity of the coffee-making device over the old-fashioned espresso pot, which required you to be the robot by grinding the beans, pouring in the water, adding the grounds to the pot, putting it on the stove to boil, and finally pouring the coffee into the cup.

A major consequence of the "improved," high-tech (read: high-complexity) coffeemaker is that you are no longer able to maintain the machine. If something goes wrong with its microprocessor "brain," or with the supply of water, or with the high-pressure pump, or God knows what else, you're out of luck. The system has collapsed, and you're powerless to fix it. And good luck in getting a human to answer your call to the customer service line.

Of course, a complexity overload in your coffeemaker is just a nuisance. Such an overload in your car is a much bigger problem. And when one occurs in an infrastructure you rely upon for your daily life needs, things really start getting serious.

In a 2005 memo to software developers at Microsoft, Ray Ozzie, former chief technical officer at the firm, wrote, "Complexity kills. It sucks the life out of developers, it makes products difficult to plan, build and test, it introduces security challenges, and it causes end-user and administrator frustration." He went on to suggest ways that might help keep complexity under control.

Ozzie typed those words at a time when Windows 2000 contained about thirty million lines of code. Its successor, Windows XP, upped the ante to forty-five million lines, and while Microsoft has wisely refrained from announcing the number of lines of code in Windows 7, it must surely be well over the fifty-million mark. But so what? Even if Microsoft could rein in the size (read: complexity) of the operating system, add-ons from developers, browser plug-ins, wikis, and the like would push the lines of code lurking inside your computer into the hundreds of millions. The point is that computing systems are not designed, they evolve. And as they evolve they grow beyond the bounds of our ability to fully control—or even fully understand—them. In some ways, these systems literally take on a life of their own. Here we come to one of the major lessons of this book: the lives of these complex systems don't remain static forever.

Anyone with a 401(k) plan should have come up against this last point in 2008. In an earlier, more leisurely bygone era, banks went into insolvency when borrowers didn't repay their loans held by the bank. But in today's world, it is not the loans the banks hold but the complexity of their assets that can send them over the edge. The infamous Lehman Brothers bankruptcy in 2007 is a prima facie case in point. The bank fell into an ultimately fatal liquidity crisis because it could not prove to the market that its assets were solid. In short, Lehman's didn't have a clear enough picture of how sound its assets were for the simple reason that it didn't have a clue as to how to value and rate the risk of those assets. Basically, the assets were out of contact with economic fundamentals and took on a life of their own.

Complexity is the culprit here. When bank assets are so complex that nobody inside or outside the bank can understand them, investors refuse to supply cash to pump up the bank's liquidity. As a result, banks hold on to the money they do have and stop loaning it to customers. But when the credit markets freeze, so does a capitalist economy since the lifeblood of such an economy is a reliable, ongoing supply of credit.

As we'll see in the next section, the problem here is that the entire financial system has become far too complex to be sustained. There is a case to be made that we have reached a state of institutional complexity that is impossible to simplify, short of a total collapse. The world's biggest banks need to become simpler, a lot simpler. But it's almost impossible for complex, bureaucratic, and publicly traded organizations like Citibank or UBS to voluntarily downsize. What happens when they don't or can't is a major part of the central puzzle I unravel throughout the rest of this book.

ARE WE DOOMED?

Joseph Tainter is an archaeologist at Utah State University, who has spent most of his career studying ancient societies. He is deeply worried that the ever-increasing complexity of the modern world will ultimately be our undoing. (He's summarized his ideas in this regard in the already-classic 1988 book, *The Collapse of Complex Societies*.) Tainter argues that as humans left the hunter-gatherer state and started settling down in communities, we have had to solve an increasing number of problems to overcome the vagaries of nature and enable survival of a large, clustered population. As different levels of organization are added, like the management structure of a city, a tax authority to collect resources to support that structure, a defense establishment, and the like, there is a cost to be paid for each successive layer. Tainter shows convincingly that the common currency for all these costs is energy, and that the increasing complexity of the system generated by adding these layers leads to the law of diminishing returns: more and more energy spent for less and less additional benefit.

For instance, each extra dollar invested today in research gives rise to fewer patents. The crunch comes when we recognize that societies must continually solve problems in order to keep growing. But the solution to these problems requires ever more complex structures. Ultimately, a point is reached where all the resources of the society are consumed just in maintaining the system at its current level. At this point, the society is experiencing a complexity overload; no further degrees of freedom exist for coping with new problems. When the next problem appears, the system cannot accommodate it by adding more complexity. So it collapses quickly through an X-event that rapidly reduces the complexity overload. Sometimes the X-event takes the form of a financial calamity or a political revolution, but often enough in history it's been a war, big or small, civil or otherwise, that clears out the accumulated bloat of complexity. Afterward the society can then rebuild from a much lower level. The well-chronicled "decline and fall" of the Roman Empire is but one of many such examples.

Of course, one might argue that innovation, technological advances, and as-yet-unimagined scientific discoveries will enable us to circumvent this "law of increasing complexity." No one can say for sure. But the facts of the way complex systems work, not to mention the historical record, argue against it. Ultimately, the question is, For how long?

In detailed studies of social organizations, system scientists have discovered that as the complexity of an organization increases, more levels of management are introduced in a hierarchical fashion. But in a hierarchy there has to be some "leader" who surveys and comprehends the entire structure and can broadcast instructions to those lower down the social totem pole. The complexity eventually becomes too great for this process to function, at which point hierarchies give way to decentralized structures with decision making distributed among many people. This seems to be the precarious perch upon which societies rest today.

At first glance, one might imagine that such a decentralized structure is more robust against unplanned-for disturbances than the hierarchical pattern. But as political scientist Thomas Homer-Dixon notes in his book *The Upside of Down,* "Initially, increasing connectedness and diversity helps, but as the connections become increasingly dense, the system gets very strongly coupled so that a failure in one part reverberates throughout the entire network." As Homer-Dixon observes, "The intricate networks that tightly connect us together— and through which people, materials, information, money and energy move—amplify and transmit any shock." So even a seemingly minor glitch in the fabric of society, such as a terrorist attack, a financial failure, or the outbreak of a disease can destabilize the entire edifice.

What to do?

The brute-force solution to the problem of increasing complexity is to simply reduce the complexity of society by moving back to an earlier style of life. Tainter says he knows of only one society in history that ever voluntarily took this huge step down the ladder of complexity. That was the Byzantine Empire, which simplified itself after losing most of its territories to the Arabs. In this case, cities vanished, the Byzantine economy became a lot less "byzantine," and the professional army gave way to local militias. While some commentators argue even today for such a return to a simpler way of life, it doesn't seem an idea that's likely to catch on. People's lives in modern society are too intricately intertwined with various infrastructures—food and water supply, energy supply, transport, communications, and others—to withdraw from the "drug" of modernity without unacceptably painful withdrawal symptoms. Almost no one wants to pay that price.

The only realistic alternative on offer is to start loosening up the tightly bound interconnections brought on by phenomena such as globalization. People will have to recognize that redundancy in a system is not necessarily a bad thing, and that maximizing the efficiency of a system simply for the sake of squeezing out the greatest possible profit is short-term, tunnel-vision thinking. After all, how useful are those profits when they cause the system itself to collapse?

I leave this story to be told in more detail later. My goal at the moment is to outline the deep complexity factors underpinning the entire social enterprise. Sustainability is a delicate balancing act calling upon us to remain on the narrow path between organization and chaos, simplicity and complexity. Following a tour through a world of potential disasters, catastrophes, and extinctions waiting to strike us down (Part II of the book), I'll return to a deeper consideration of these matters of sheer survival and how it may be achieved without having to resort to radical downsizing (Part III).

Before moving to those parts, though, let's take a longer look at various ways that complexity can manifest itself in the social structures we are all embedded within. We'll see that the term "complexity" is not monolithic; rather, it has many sides. It's important to have an idea of what these are if we want to see how to reduce complexity to a manageable level for social sustainability without throwing out the baby with the bathwater.

SEVEN FACES OF COMPLEXITY

WHEN I FIRST WENT TO THE SANTA FE INSTITUTE ABOUT TWENTY years ago, one of the bright postdocs on the staff was Seth Lloyd, now a distinguished professor at MIT and a leading light in the quantum computing community. At the time, complexity science was just starting to gain traction in the intellectual world, and a lot of people both inside and outside academia were wondering just what it meant to say a system is "complex" (as opposed, for instance, to being just "complicated"). As academics are fond of trying to nail down the terms of an argument before jumping into a debate, the first line of attack on the question was to attempt to formulate some sort of definition of the complexity of a system, together with a scheme for measuring exactly how complex any given system actually is. The misguided hope, of course, was twofold: (a) to dream up a unit of measure of complexity that would be universally accepted by the complexity science community, and (b) would enable us to generate a number characterizing the complexity of any given system, thus allowing one to say something like the Internet is 3.141592 times as complex as, say, the US postal system. From what's been learned about complex systems during the intervening decades, this exercise seems hopelessly naive, if not just plain silly. And it's equally as unclear today as it was then whether any such magic measure would tell you anything truly useful about a system. But such are the blind alleys and culs-de-sac one wanders down during the fumble-and-stumble stage of any new intellectual enterprise.

In an effort to sort out the various ideas about complexity floating about at the time, Lloyd took on the task of sifting through the literature on measures of complexity and grouping them by whether they focused on the structure of the system, its behavior, the perception of observers about the way the system works, and other such categories. In the end, he put together what I believe is a still-unpublished account of what he'd discovered, titling it "31 Flavors of Complexity" in honor of the well-known slogan employed by the American ice cream chain Baskin-Robbins.

As I noted a moment ago, many are the faces of the mask of complexity. What is and isn't complex depends to a large degree not only on a target system, but also on the system(s) the target interacts with, together with the overall context in which the interacting systems are embedded. So while it might be nice to have uniform agreement, perhaps even backed by a single number, as to the complexity of a particular system, I have to disabuse the reader of expecting to find that magic elixir here. Instead, what I want to show now is what might be called the "seven faces of complexity." Each of these faces will appear as the dominant face in one or another of the various situations I describe in Part II of the book. So to set the scene, let's look under the hood more closely at the various ways complexity shows up in the real world.

COMPLEXITY PRINCIPLE I: EMERGENCE

A collection of individuals in interaction form a "system." And this system *as a whole* often possesses its own unique properties that do not exist at the level of the individuals themselves. Such *emergent* characteristics are termed "systemic" traits, as opposed to those of the individuals composing the system, since they arise from the *interactions* of the individuals. Good examples of emergent behaviors or traits are congestion on a freeway, points scored in a football game, or a price change in a financial market. Thus one car leaving a freeway causes no congestion at all. Hundreds of cars trying to get off at the same exit to reach the stadium before kickoff time creates a jam. Likewise, no single football player can determine the outcome of a football play, regardless of how great he might perform. It takes the interaction of a group of players to determine whether the play ends up in a touchdown or not. So points scored is an emergent aspect of the game. In a similar manner, the interaction of the decisions by traders in a financial market to buy, sell, or hold sends the price up or down. That price change is also an emergent phenomenon, one that is determined both by the decisions of the traders and their interaction.

Emergent traits and/or behaviors are frequently seen as "unexpected" or "surprising." This is because we generally know something about the characteristics of the individual objects, but not about the overall total systemic properties arising from the interactions of those objects. We know about the behavioral patterns of individual rioters on the streets of London or Cairo, for instance, but we don't know how the behaviors of those individuals interact to create even a civil disturbance, let alone an X-event like the overthrow of a government. That disturbance is an emergent property that's characteristic of the total system, not something to be found in the behavioral makeup or actions of any particular rioter.

COMPLEXITY PRINCIPLE II: THE RED QUEEN HYPOTHESIS

In Lewis Carroll's classic work *Through the Looking Glass,* the Red Queen remarks to Alice, "In this place it takes all the running you can do to keep in the same place." This idea was brought to the world of science by ecologist Leigh van Valen in 1973, who noted that in any system composed of a collection of adaptive, evolving organisms, each member must evolve to meet the competition from the others in order to avoid extinction. In short, you have to evolve as fast as you can just to stay in the game. A consequence of this principle is that the system's overall fitness tends to move higher and higher up the complexity ladder—until it doesn't! At that point a systemic collapse takes place, generally through the agency of yet another system that outcompetes the first one. (Here there are echoes of Joseph Tainter's claims discussed earlier regarding the increased complexity of a society ultimately leading to its undoing.)

An excellent example of this principle is one of the most publicly visible aspects of the phenomenon of globalization, namely, the loss of manufacturing jobs in the United States to China and elsewhere in Southeast Asia. Consider the situation with China. Here we have two systems in interaction, the manufacturing sectors of each country involved. One of them, the manufacturing sector in the United States, continually increased its complexity by adding layers to the structure, things like minimum-wage laws, health and safety standards, union-ization, and the like. The competing sector in China added little, if any complexity of this sort, other than modernization of its facilities and increased automation. Eventually, the complexity gap grew too large to be sustained and the X-event of a massive transfer of manufacturing jobs from one country to the other was the result. In other words, the high-complexity system was downsized by an X-event, because it failed to voluntarily downsize itself. This was truly a shock to both systems—but in very different ways.

COMPLEXITY PRINCIPLE III: NO FREE LUNCH

If you want a system—economic, social, political, or otherwise—to operate at a high level of efficiency, then you have to optimize its operation in such a way that its resilience is dramatically reduced to unknown—and possibly unknowable—shocks and/or changes in its operating environment. In other words, there is an inescapable price to be paid in efficiency in order to gain the benefits of adaptability and survivability in a highly uncertain environment. There is no escape clause!

A rock climber, for instance, could choose to attempt a solo ascent up an exposed mountain face. He might make it more often than not, but it only takes one unexpected glitch—his grip slips, a rock breaks beneath his feet, falling ice strikes him from above—to send him falling to his death. This is why most experienced climbers work in teams and take time to place a redundant network of protection as they ascend. It may be slightly less efficient in terms of the time it takes to get to the top, but if the unexpected happens, the climbers can gather themselves and still continue to the summit.

At the corporate level, there often is a trade-off between having a very specialized product line or a huge inventory of different products. As an example, think of the deep stock of books at Amazon, and compare it with a firm selling a large number of units of a single product, such as McIlhenny's Tabasco Sauce. The trade-off we see here is between operating at a high level of efficiency by focusing on *process innovation,* as at McIlhenny & Co., which improves its production process and thereby increases productivity. Or a firm may fortify its robustness to unanticipated shocks through *product diversification,* which spreads the risk and reward over a much broader line of offerings, as at Amazon.

The Amazon-style operation spreads the risk among many products so the company's eggs are not all in one basket, as would be the case for instance if Amazon focused on selling a particular volume by a name-brand author. Instead, it has "portfolio diversification," which costs money in order to provide the ability to supply just about any book in print. On the other hand, McIlhenny has optimized its operation to produce one product. So it reaps a large return on that investment in efficiency—provided Tabasco sauce doesn't go out of style. If it does, so does McIlhenny.

Thus, the low-complexity system, McIlhenny, can remain viable only if the high-complexity operation, Amazon, doesn't decide to go into the Tabasco sauce business itself via what we called above "product innovation." If it does, the complexity gap between the two systems may become too large to sustain, in which case either McIlhenny would have to upgrade its complexity via diversification of its own product line, or suffer the X-event of going out of business.

The first path of increasing complexity via innovation often faces limits as to how much complexity can be added or reduced in a given system. This is because if you change the complexity level in one place, a compensating change in the opposite direction generally occurs somewhere else. This is one of the primary reasons why it is so difficult to voluntarily reduce complexity in a big bureaucratic organization, since "complexity creep" has infested so many parts of the system that it's not possible to do much to reduce it through spot changes here and there. It's just like an old car that cannot be repaired any longer. You have to throw it away and build or buy a new one.

COMPLEXITY PRINCIPLE IV: THE GOLDILOCKS PRINCIPLE

Systems operate in the most open, dynamic, and adaptive fashion when the degrees of freedom available for them to act are like Goldilocks's porridge, not too hot and not too cold, but just right. In systems jargon, this is often termed the "edge of chaos," the fine line between when the system is too frozen in place with too few degrees of freedom to explore new regimes of behavior, and when there are so many degrees of freedom that the system is essentially chaotic and just about anything goes. The middle of the road is the right place to be to exploit the structure present, but still have enough room to move to new structures as timing and changing circumstances dictate.

Managers of national economies, for instance, must walk a thin line between permitting the freedom needed for innovation and entrepreneurship, on the one hand, and employing just enough regulation to keep malfeasance and skirting the law in check. Too little oversight results in chaos, destabilizing the system. But too much government planning and control is apt to stifle growth of the economic system as a whole.

The recent book *Red Capitalism: The Fragile Financial Foundation of China's Extraordinary Rise,* by Carl E. Walter and Fraser Howie, notes that the Chinese government reformed and modeled its financial system during the last thirty years in a very special way. The result of this structuring is that stability of the system can be maintained only behind the walls of a nonconvertible currency, a myriad of off-balance-sheet arrangements with nonpublic state entities and the strong support of its best borrowers—the politically potent National Champions, who are the greatest beneficiaries of the financial status quo. China's financial system, as such, is not a model for the West, whose populations demand a much higher level of transparency, and indeed it is not even a sustainable arrangement for China itself as it increasingly seeks to assert its influence on the world stage.

On this point, the *Economist* magazine noted that for China, "The bigger problem, though, is that the system trades almost entirely with itself. Critical information about liabilities and pricing is deliberately concealed or impossible to discern; there are no outside entities establishing the prices of goods and services by bidding in the market. That undermines efficient capital allocation and allows excesses to fester."

So what we see at work here is a situation in which to maintain a viable capital allocation process and to plug into the world economy as a fully participating country, China has walked a narrow line between a too restrictive, essentially closed, banking system and a system that is so open it leaks like the sieve that many Western banking systems, especially those in the United States and Europe, have become today.

COMPLEXITY PRINCIPLE V: UNDECIDABILITY/ INCOMPLETENESS

Rational argumentation alone is not sufficient to settle every possible claim about the occurrence or nonoccurrence of an action or behavior. Put another way, events will always occur that cannot be foreseen by following a chain of logical deductive reasoning. Successful prediction requires intuitive leaps and/or information that is not part of the original data available.

In 1931, Austrian logician Kurt Gödel proved that perfectly innocent-looking assertions exist about the relationship among numbers whose truth or falsity cannot be settled by logical deduction alone. He went on to show that such an undecidable statement is actually true; you just cannot prove it's true by using the assumptions built into the usual rules you use to generate proofs. The system is just not "strong enough" to do the job. In short, it is incomplete. And this is true for whatever logical framework you care to use; that set of rules will always have at least one such undecidable proposition. In fact, it has been shown that almost *every* statement you make about numbers is in this category. Thus, the rarities are statements that *can* be settled by a deductive argument, not those that cannot be formally proved or disproved.

So if logical systems of deduction cannot even settle questions about numbers, imagine the challenges of predicting human events. Since a level of complexity can be attached to any such statement, we can loosely restate Gödel's incompleteness theorem as: "There exist statements that are too complex for the human mind to comprehend." How does this excursion into the stratosphere of mathematics and logic connect with real-world concerns about X-events?

In 2011, the government of Hosni Mubarak in Egypt unraveled in Tahrir Square in Cairo as a follow-on to the ouster of the thirty-year Ben-

Ali regime in Tunisia a few weeks earlier. Within hours of Mubarak's over-throw, bloggers were trumpeting their wise thoughts to the world about why these events were taking place now, and even more self-importantly, how they had foreseen the changes months or even years earlier.

These analyses share the feature that they all consist of some chain of logic that starts with a cluster of circumstances (i.e., axioms) and leads inexorably (via rational argumentation) to the events that played out in Cairo, Tunis, and Damascus. Looking back a couple of decades, we see the same pattern of ex post facto logic to explain the collapse of the USSR, the unhappy ending of the war in Vietnam, and, of course, that perennial standby, the fall of the Roman Empire—shades of Nassim Taleb's remark that forecasting X-events is wrapped up with the human proclivity for telling ourselves Just-So stories that appear to make the event's occurrence a foregone conclusion. No doubt about it, political commentators and historians love the seeming inevitability of after-the-fact logic to support their so-called explanations.

But these sorts of explanations are a dime a dozen. The real, rather than fool's, gold rests with ex ante, not ex post facto, logic; forecasting, not explaining, via a chain of rational arguments that such events are at least very likely, if not inevitable. This is doubly true in the case of X-events, since they are the real game changers that turn whole societies upside down. Using Gödel's arguments in a metaphorical, if not mathematical, sense, the chain of chance is at least as important as the chain of logic in identifying what is or is not likely to unfold in these social domains. In short, you can't get it all by rational thought alone. And, in fact, what you usually cannot achieve is precisely what you most want: a clear, unambiguous picture of an impending X-event.

COMPLEXITY PRINCIPLE VI: THE BUTTERFLY EFFECT

While studying mathematical models of atmospheric processes back in the 1970s, MIT meteorologist Ed Lorenz discovered one of the signature features of a complex system: a seemingly insignificant change or disturbance in one part of the system can cascade through the entire network to produce a major change in another part and/ or at another time. Lorenz called this the "butterfly effect," referring to a process by which a butterfly flapping its wings in the Brazilian jungle today might give rise to a hurricane in the Gulf of Mexico next week. The basic idea is that complex systems can display behavior that might be pathologically sensitive to seemingly minor changes in the starting state of the system. Here is an example showing this property in spades.

In early 2000, Theresa LePore was designing the ballot that Palm Beach, Florida, voters would use in the US presidential election to be held in November of that year. She decided to make the typeface larger on the ballot, presumably so that the many octogenarian voters in Palm Beach would find it easier to read. For reasons unknown, she didn't realize that this change would turn the ballot into a two-page document instead of one, which might lead to confusion for the voters as to which button to press on the voting machine.

When the votes were counted, 19,120 voters hit the buttons for both Pat Buchanan and Al Gore, resulting in their ballots being thrown out. Further, more than three thousand people voted for Pat Buchanan, who had expected to receive only a couple of hundred votes from this community. Presumably, most of the extra votes were really intended for Gore but, through confusion about the ballot, went to Buchanan instead. The end result of all this was that around twenty-two thousand votes for Gore were disallowed. If they had been credited to him, Florida would have gone to Gore and he would have become the forty-third president of the United States instead of George W. Bush. Many have since argued that if this had occurred, the world would be a very different place today. So Ms. LePore's failure to note that her change in the format of the Palm Beach ballot might confuse voters rather than help them can be seen as a flapping butterfly wing that changed the course of modern history.

COMPLEXITY PRINCIPLE VII: THE LAW OF REQUISITE VARIETY

Now we come to the most important complexity principle of all, at least for the purposes of this book. This is the principle explaining why an X-event typically arises as the means of closing an unsustainable gap in complexity levels between two or more systems in interaction.

In the 1950s, cyberneticist W. Ross Ashby had an "aha!" insight: the *variety* in a regulatory system has to be at least as great as the variety of the system it's supposed to regulate if it's to do its job. Here by "variety" he meant the number of degrees of freedom each system has at its disposal to act at any given moment. For the purposes of the stories I tell in this book, the terms "variety" and "complexity" can be used more or less interchangeably. In contemporary terminology, we'd understand Ashby's law as saying that the control system has to be at least as *complex* as the system to be controlled; if not, the complexity gap between the two can—and often does—lead to all sorts of unpleasantries.

Greek business consultant Alexander Athanassoulas has given a very interesting illustration of Ashby's law in the context of tax evasion, a topic of increasing concern in debt-ridden countries around the world. As each year goes by, countries pass bodies of legislation to rein in and penalize tax evasion. But the variety of actions available to the tax collectors will never be matched by the bewildering array of tools employed by accountants, lawyers, and tax evaders to avoid paying their share of the national financial burden. In short, the variety of the general population ready to evade taxes can never be matched by the vastly smaller variety of tools available to the tax regulators (inspectors). This means that tools are needed to reduce the variety on the side of the evaders rather than trying to control tax evasion after the fact. Athanassoulas suggests solutions like a reduction of tax rates, a more equitable range of tax rates across the population, and other vehicles of this sort as means to reduce the variety of the evaders.

• • •

WE'VE NOW SEEN SEVEN FACETS OF COMPLEXITY AND HOW THEY EACH give rise in their own characteristic way to X-events of various sorts. The table below summarizes these manifestations of complexity with a catchphrase characterizing the way an X-event might arise via this route. It's worth mentioning that this list is neither exhaustive nor are the items mutually exclusive. Any given X-event may well be generated by a combination of several of these complexity principles. Generally speaking, though, there is one dominant principle, the others playing the role of supporting characters in the drama of the X-event. What matters is that the X-event itself ultimately stems from *complexity running amok*.

SEVEN PRINCIPLES OF COMPLEXITY
AND THEIR PROPERTIES

Complexity Principle	Property
Emergence	Whole is different from the sum of the parts
Red Queen Hypothesis	Evolve to survive
No Free Lunch	Tradeoff between efficiency and resilience
Goldilocks Principle	Degrees of freedom "just right"
Incompleteness	Logic alone is not enough
Butterfly Effect	Small changes can generate huge effects
Law of Requisite Variety	Only complexity can control complexity

The line of argument I pursue in the pages to follow is that Complexity Principle VII, the law of requisite variety, is just a bit more equal than the other principles when it comes to the occurrence of an X-event. As with natural systems, human systems too seem to work best when all the various subsystems making up a society are in some measure of balance and harmony. When the respective complexities of these subsystems deviate too much from one another so that complexity mismatches, or "gaps," arise, the system necessarily tries to reconfigure itself so as to narrow or balance out these gaps. Since politicians, business leaders, and the general public do not normally see it in their best short-term interests to take an immediate loss in exchange for actions that place the social system in a better position to survive over the long haul, it's often the case that the natural dynamics of the system have to step in to rectify a complexity imbalance. Those self-organized, systemic actions are usually quick and disruptive, often involving the appearance of an X-event, or two or three, to get people's attention and close a widening gap as quickly as possible.

I now proceed to Part II, which gives a more extended account of eleven different human-caused X-events, all of which has occurred at some time in the past and can easily send human life back into a horse-and-buggy era should it take place again.

PART II

GETTING DOWN TO CASES

X-EVENTS COME IN MANY SIZES, SHAPES, AND FORMS. CONSIDER THE year 2003, when several front-page extreme events took place. Among them were a huge power failure in the Northeast of the USA, the outbreak of the SARS virus, and a magnitude 9.1 earthquake and consequent tsunami in Sumatra. Yet the number of people directly affected was dramatically different in each case. The power failure impacted 55 million people but resulted in very little loss of life, while 8,273 confirmed deaths were reported from the SARS outbreak. But the 283,106 fatalities from the Indonesian quake dwarfed the number of deaths attributable to either of the other two X-events. So if you measure the magnitude of an X-event by its lethality, the Sumatran earthquake was the worst, hands down. But if you measure things on the scale of damages, both material and financial, it may be another story. Yet a third angle is the psychological damage that accompanies the loss of homes and jobs, not to mention the anxiety and uncertainty of wondering when the event will actually end. The point to emphasize here is that the "X-ness" can vary widely. So to really understand these events, we cannot continue to speak in broad generalities.

Understanding which types of extreme events can be anticipated, what sorts can only be endured and hopefully recovered from and those we can only wish, pray, and hope never occur, requires a deeper look into a broad spectrum of scenarios. That is the primary goal of this part of the book.

In the pages that follow, I present eleven minichapters, each of which tells the story of a particular X-event. I've chosen these examples to run the gamut from the relatively familiar (the peak oil crisis) to the seemingly offbeat (an earthly implosion via the creation of exotic elementary particles). In this catalog of catastrophes, I've deliberately avoided "natural" X-events like volcanoes, asteroid impacts, or even global warming not because they are any less cataclysmic or any less likely to send humankind back to a more primitive way of life, but because they are by now representative of events that have been so extensively chronicled that one would hardly even term them "surprises" any longer. So for the sake of novelty, if nothing else, I like my list better by not including these particular sorts of garden-variety X-events.

But novelty alone is a thin reed upon which to rest the stories told here. The real foundation is the way "complexity overload" enters into the occurrence of each and every X-event recounted in the following pages. In each chapter, the reader will find one or more of the complexity principles outlined in Part I sitting at the very core of the cause of that chapter's defining X-event.

As WE NOTED EARLIER, NOT ALL EXTREME EVENTS ARE CREATED equal. And in a 2004 article, British engineer C. M. Hempsell introduced three categories of X-events:

1. *Extinction-Level Events*: An event so devastating that more than a quarter of all life on Earth is killed and major species extinction takes place. *Example*: The end of the Cretaceous period when around 80 percent of all existing species vanished.
2. *Global Catastrophes*: An event where more than a quarter of the world's human population dies. *Example*: The Black Death of the Middle Ages.
3. *Global Disasters*: Events in which a small percentage of the population dies. *Example*: The Spanish influenza epidemic of 1918.

Thus we see that the terms "extinction," "catastrophe," and "disaster" refer mostly to the *intensity* of the event (its impact magnitude), not to the *timescale* of its occurrence (the unfolding time) or how long the impact lasts (its impact time). Finally, there is the *likelihood* of the event's occurrence, a matter left untouched in Hempsell's taxonomy.

All these factors are vitally important when it comes to consideration of how seriously to take the idea of preparing for such events. So I need to say a few more words about timescale, timing, and the likelihood of these game-changing X-events. First, timescale.

Some types of events simply take time to wreak their havoc. A worldwide plague, for instance, cannot infect everyone in a few hours. Even the most contagious disease requires a transmission process that takes many weeks to make its way around the planet—today's fast-paced, jet-setting world notwithstanding. On the other hand, an out-of-the-blue asteroid strike can do at least the blast damage phase of its handiwork more or less instantaneously.

Turning to the issue of timing as opposed to timescale, we ask *when* an event will occur. Note that this is different from asking how likely it is that a given event will take place. The timing refers to whether there are preconditions to the event happening that relegate it to sometime in the distant future, if at all, or if it might take place at any moment. The range of possible answers then runs from immediately to never.

To illustrate this range, let's look again at the nanoplague extinction event, the gray goo problem. We ask, When might this X-event happen? At present, nanotechnology has not yet reached a state where self-replicating nanobots are possible. But most everyone in the nanotech business agrees that there are no logical or physical obstacles to this development. The technology is just not there—yet! So the timing of a nanocancer would be set by when this purely technical barrier is surmounted. The consensus seems to be that the right answer is a few years, perhaps a decade or so at the most.

On the other hand, the timing for an event like a hostile alien invasion might be set at already, now, never, or any time in between. There is simply not a shred of hard, incontrovertible evidence for helping us choose one of these times over another.

A somewhat more interesting case is the timing of something like a supervolcano eruption in Yellowstone National Park. Geophysicists and vulcanologists know that the entire park is the caldera of an ancient volcano that last exploded about 650,000 years ago. Evidence suggests that this will happen again. When? Nobody can really say. But it is almost certainly neither now nor never. But this is just one such caldera. The earth is covered with many others of a similar type, and all it would take to snuff out a majority of planetary life is for one of these to blow its top. So the question of timing comes down to when any one of these volcanoes will erupt, not just Yellowstone. This moves the timing down the scale, but probably still in the range of many centuries if not millennia.

Finally, let's consider the big one: likelihood.

How likely is it that we will be destroyed by an alien invasion, an event that has never occurred and for which we have absolutely no information suggesting that such an event even could occur? Or how likely is it that the caldera forming Yellowstone National Park will rumble back to life? These are two types of X-events, one for which no database of any kind exists, the other for which there is definite evidence of past occurrences. They illustrate the problem of trying to employ standard statistical and probabilistic tools for estimating the likelihood of an X-event taking place. In both cases, humanity may be sent back to the Stone Age, if not extinguished altogether, by the event's occurrence. But in the first case we can only speculate (i.e., guess) at the likelihood of the event's occurrence, while in the second case we can at least try to employ tools of extreme-event analysis to come up with something approaching a meaningful number for the event's "probability."

Note here that when I talk about likelihood, I'm not referring to "likelihood within a given period of time." The actual timing is already built into the analysis I gave earlier. So for likelihood, I mean "at any time," or put another way, an answer to the question "What is the likelihood of the event *ever* taking place?" This is a strong condition and essentially rules out the answer never, since something that is not absolutely excluded by logic or physics has to be given at least a minimal chance of happening at some time. But, again, not all events are created equal. And some, like a killer earthquake, are far more likely than others, such as Earth being fried by gamma rays from a hyper supernova on the other side of the Milky Way galaxy.

Putting all these thoughts together, for the sake of classification I'll divide likelihood into five categories:

Virtually Certain: Scenarios that will almost surely take place, such as an asteroid impact or a major earthquake or a financial crash. These are events that have happened many times before, and for which we have ample evidence in the geological and historical record to believe that they will certainly happen again.

Definitely Possible: Scenarios that have either happened before or that we have evidence suggesting they may be in the process of already unfolding. This group includes things like a pandemic, a global nuclear holocaust, a runaway Ice Age, or destruction of Earth's ozone layer.

Unlikely: Events for which we have no record of past occurrence, and that while possible are by no means required to take place. A nanocancer or a massive cultural decline falls into this category.

Very Remote: Events that are so unlikely as to have virtually no possibility of impacting humanity at any time. The likelihood of Earth being "reconfigured" by some time traveler who steps on an ancient mammal that just happens to be the original ancestor of the human race illustrates events of this category.

Impossible to Say: These are occurrences for which we have absolutely no idea at all regarding their likelihood. An invasion by hostile aliens, or a takeover of human civilization by intelligent robots are good examples here.

I've now divided humanity-threatening events along three dimensions: *timescale,* measuring how long it may take for the X-event's damage to fully unfold; the *timing* as to when the X-event may occur; and the *likelihood* of the event, characterizing its chances of ever taking place.

So far I've said little about the actual cause(s) for a human-created X-event. It's useful then to take a brief, high-level overview of the entire landscape and consider the question of whether it's inevitable that one or another of the disasters, catastrophes, or extinctions I discuss in the book will ultimately do us in. In particular, my focus will now shift almost exclusively to human-generated, or at least human-abetted, X-events, since the cause of X-events arising from nature are by now rather more well understood than what we humans ourselves can cook up. Of course, this does not mean that we really have a good handle on the game changers that nature throws our way either, just a bit better understanding than for the human-caused ones.

Before proceeding to the chapters themselves, I want to emphasize that the X-events that play the starring role in the stories to follow are not fiction, science- or otherwise. The majority of them have happened before, and it wouldn't take much imagination to see how they could happen again. So don't be seduced simply by the story lines, compelling as they are. Keep in mind the ways human activity can act to cause, or at least exacerbate, the X-event. Also keep in mind that it's not just at the level of the individual that we are our own worst enemy. The same principle works at the societal level as well, as these stories graphically illustrate.

DIGITAL DARKNESS

A LONG-TERM, WIDESPREAD OUTAGE OF THE INTERNET

BAD SIGNALS

IN THE SUMMER OF 2005, COMPUTER SECURITY CONSULTANT DAN Kaminsky was at home recovering from an accident. While recuperating in a pain killer–induced haze, he began thinking about some Internet security issues he'd pondered earlier, specifically focusing his deliberations on the Domain Name Server (DNS) component of the Net. This is the part that serves as a dictionary for translating everyday-language domain names like *oecd.org* or *amazon.com* into the twelve-digit Internet Protocol (IP) addresses that the system actually understands and uses to route traffic from one server to another. For some time Kaminsky had felt that there was something not quite right about the DNS system, that somewhere in it there was a lurking security hole that had existed since the system was put in place in 1983, a hole that could be exploited by a clever hacker to gain access to virtually every computer in the entire network. But he could never quite put his finger on what, exactly, the problem might be.

Then in January 2008, Kaminsky hit on the answer. He tricked the DNS server of his Internet provider into thinking he knew the location of some nonexistent pages at a major US corporation. Once the server accepted the bogus page made up by Kaminsky as being legitimate, it was ready to accept general information about the company's Internet domain from him. In effect, Kaminsky had found a way to "hypnotize" the DNS system into regarding him as an authoritative source for general information about any domain name on the *entire* Internet. The system was now ready to accept any information he wanted to supply about the location of any server on the Net.

Kaminsky immediately recognized that he had just entered hacker's heaven. What he found was not simply a security gap in Windows or a bug in some particular server. Instead his discovery was an error built into the very core of the Internet itself. He could reassign any web address, reroute anyone's e-mails, take over bank accounts, or even scramble the entire Internet. What to do? Should he try it? Should he drain billions out of bank accounts and run off to Brazil? It's difficult to imagine being faced with this kind of power over the lives of billions of people worldwide. Maybe he should just turn off his computer and forget it. If what he found was reported in even one blog or website, within seconds unscrupulous hackers around the world would pounce upon it and possibly do irreparable damage to the global economy and to everyone's way of life.

What Kaminsky actually did was to contact a handful of web security gurus, who then arranged to meet in emergency session as a secret geek A-team. At this meeting, they created a temporary fix for the hole Kaminsky had found to punch his way into the DNS system. But as he put it in concluding an address about the problem given on August 6, 2008, at a hackers' convention in Las Vegas, "There is no saving the Internet. There is [only] postponing the inevitable for a little longer."

So it remains to this day. And this is not a Hollywood fantasy scenario either, as a single individual "playing around" in his or her garage has as good a chance to take down a chunk of the Internet as a team of computer specialists at a government security agency. In this game, inspiration and ingenuity are at least as likely to strike a lone individual as it is to descend upon a group.

Kaminsky's discovery of a hidden flaw at the very heart of the Internet brings into focus the threat posed by a massive Internet failure to our twenty-first-century lifestyle. From e-banking to e-mail to e-books to iPads and iPods, and on to the supply of electricity, food, water, air and surface transport, and communication—*every* element of life as we know it today in the industrialized world is critically dependent on the communication functions provided by the Internet. When it goes down, so does our way of life. So when we start talking about a massive failure of the Internet, the stakes are about as high as they can get. And as Kaminsky's hack showed vividly, this system is far from being immune to a catastrophic breakdown.

Since Kaminsky's discovery struck at the very core of the Internet, perhaps this is a good time to say a few words about how it was created, as well as a bit about what people were thinking in those days more than half a century ago. Ironically, the system in question was developed to help some of us *survive* an X-event.

The origin of the Internet dates back to the 1960s when the US government began collaboration with private industry to create a robust, fault-tolerant distributed computer network. What the government wanted was a network that was not rooted at one spatial location and thus would be able to continue functioning even when many of its nodes and/or links were destroyed, temporarily broken, or otherwise out of service. It should come as no surprise to learn that the Cold War mentality of the time was the big motivator for creation of what became the Internet, as the US defense establishment needed a command-and-control system that would remain operational even in the face of an X-event: a full-scale nuclear attack by the USSR.

The communication system originally put into service was termed "ARPAnet," named for the Advanced Research Projects Agency (ARPA), a blue-sky research arm of the US Defense Department. Commercialization took place in the 1980s, along with the appellation "Internet" replacing the ARPAnet. Since then the capabilities of our communication systems have defined our business structures. Information you can quickly accumulate and process supports the economy by facilitating faster decision making, increased productivity, and thus faster economic growth. Speed and access to information determines today's customer-business relationship.

The distributed nature of the Internet is reflected in the fact that there is no centralized governing structure that "owns" the Internet, as only the two "names spaces" for the system, the Internet Protocol (IP) address space and the Domain Name System (DNS), are governed by a central body.

In essence, then, we have ended up with a communication system, used currently by approximately one-quarter of the people on the planet, that rests on 1970s-style notions of computer networking and computer hardware. Today, the Internet is being used to support services that it was never designed for, as we are converging to a situation in which all data types—voice, video, and verbal information—are being loaded onto it. With this fact as background, it's no wonder that the technological and lifestyle changes of the last fifty years are putting an ever-increasing strain on the system's ability to serve the needs of its users. A few eclectic examples will hammer home this point.

Item: In mid-October 2009, a seemingly routine maintenance of the Swedish top-level domain *.se* went badly off track when all domain names began to fail. Swedish websites could not be reached, Swedish e-mails failed, and even several days later not all systems were fully functional again. The entire Swedish Internet was broken. What went wrong? Investigations suggest that during the maintenance process, an incorrectly configured script, intended to update the .se zone, introduced an error into every .se domain name. But this was only a theory. Another possibility floated at the time was that the Swedish Internet may have crashed when millions of Japanese and Chinese men disrupted the country's Internet service providers searching for Chako Paul, a mythical lesbian village somewhere in Sweden! So by this hypothesis, the entire network was taken down by Asian men googling a seemingly nonexistent Swedish "village."

Item: In November 2009, the US television newsmagazine *60 Minutes* claimed that a two-day power outage in the Brazilian state of Espirito Santo in 2007 was triggered by hackers. The report, citing unnamed sources, said the hackers targeted a utility company's computer system. The outage affected three million people, a precursor to a huge blackout in 2009 that took out Brazil's two largest cities, São Paulo and Rio de Janeiro, along with most of the country of Paraguay. As it turned out, neither of these disruptions appear to have had anything to do with infiltration of any computer system. Rather, the 2007 failure was due to simple human error: faulty maintenance of electrical insulators that allowed them to accumulate so much soot they eventually shorted out. Explanations offered for the much larger power failure in late 2009 are far more interesting, ranging from a major storm destroying power lines at Itaipu Dam, the source of 20 percent of Brazilian electricity (meteorological records show no storm in the vicinity of the dam during the period in question), to renegade Mossad agents hacking into the national grid (the explanation favored by Brazilian president Lulu da Silva), to a "butterfly effect" stemming from the shutdown at CERN in Geneva of the Large Hadron Collider during approximately the same period as the blackout, and even to UFOs in the form of an alien mother-ship harvesting electricity from the generating station. In short, nobody knew anything!

Item: On May 17, 2007, the Estonian Defense Ministry claimed the Russian government was the most likely cause of hacker attacks targeting Estonian websites. They said more than one million computers worldwide had been used in recent weeks to attack Estonian sites following removal of a disputed Soviet statue from downtown Tallinn, the Estonian capital. Defense Ministry spokesman Madis Mikko stated, "If let's say an airport or bank or state infrastructure is attacked by a missile it's clear war. But if the same result is done by computers, then what do you call it?"

Item: China Telecom reported that according to the China Institute of Earthquake Monitoring, on December 26, 2006, between 8:26 P.M. and 8:34 P.M., Beijing time, earthquakes of magnitude 7.2 and 6.7 occurred in the South China Sea. The undersea communication cables Sina-US, Asia-Pacific Cable 1, Asia-Pacific Cable 2, FLAG Cable, Asia-Euro Cable, and FNAL cable were all severed. The location where these cables broke was about 15 kilometers south of Taiwan, severely affected international and national telecommunication in neighboring regions for weeks until repairs could be made.

Other reports at the time stated that communications directed to the Chinese mainland, Taiwan, the USA, and Europe were all seriously interrupted, and that Internet connections to countries and regions outside the Chinese mainland were very difficult. In addition, voice communication and telephone services were also affected.

These news reports were a dramatic understatement. China and Southeast Asia saw their communication capacity fall by over 90 percent, in what mainland Chinese began referring to as the "World Wide Wait." What this outage underscored was the dilapidated state of China's telecommunication technology. As the global news service AFP put it, "China is relying on 19th-century technology to fix a 21st-century problem."

Finally, a few paragraphs about something nobody believed was possible: the *total and complete* disappearance of the Internet in a major region of the world.

Item: At 12:30 A.M. on the morning of Friday, January 28, 2011, the Internet died in Egypt. At that moment, all Internet links connecting Egypt to the rest of the world went dark, not so coincidentally, as demonstrators protesting against the thirty-year rule of President Hosni Mubarak's brutal regime were gearing up for a further round of marches and speeches. Egypt had apparently done what many technologists thought was unthinkable for a country with a major Internet-based economy: it unplugged itself entirely from the Internet in an effort to stifle dissent. Leaving aside the issue of why this took place, which is not at all hard to understand, the technical aspects of *how* it happened are worth a quick look.

In a country like the United States, there are numerous Internet providers and enormous numbers of ways of connecting. In Egypt, four providers actually control almost all links to the system, and each of these operates under strict control and strong licensing from the central government. In contrast to the United States, where it might be necessary to call hundreds or even thousands of service providers in order to try to coordinate them all to throw the "kill switch" at the same moment, in Egypt this coordination problem could be solved with a few phone calls. So the reason this total blackout could happen in Egypt is that it is one of the few countries where all the central Internet connections rest in so few hands that they can all be cut at the same time. Here we see an obvious complexity mismatch between the control system of the Egyptian Internet and its users.

Experts say that what sets Egypt's action apart from those of countries like China and Iran, who have also restricted segments of the Internet to close off dissent, is that the entire country was disconnected in a coordinated effort, and that every type of device was affected, ranging from cell phones to mainframe computers. One might wonder why this hasn't happened more often in places like Iran or even the Ivory Coast, where political dissent is an ongoing irritant to the ruling authorities. The reason is largely economic. In today's world, a country's economy and markets are just too dependent on the Internet to shut it down over such an ephemeral matter as a possible regime change. Dictators come and go, but money never sleeps.

Human error (or design) is certainly the primary means by which the Internet can be compromised. But, as always, the joy is in the details. And the specifics can run through a spectrum of methods ranging from Kaminsky-style assaults on the DNS system to those aimed at the end user. Even attacks aimed at the Internet's social fabric have been suggested, such as spamming people with death threats to convince them that the Internet is unsafe to encouraging webmasters to unionize and go on strike. In short, there are many ways to bring down the system, or at least huge segments of it. The most amazing fact of all is that it hasn't happened more frequently.

These stories could easily have been multiplied severalfold. But none of them represent the kind of event that would send our global society tumbling into an abyss, even though all are disasters to one degree or another. More ominously, all could have easily been scaled up to a genuine worldwide catastrophe if events had fallen just a bit differently. The most important fact, however, is that none of the countries involved were really prepared to deal with this kind of attack on their infrastructure. As the old Viennese saying goes, these situations were desperate—but not serious. The take-home message is clear: the infrastructures humans most rely upon for just about every aspect of modern life are totally dependent on computer communication systems, the vast majority of which are linked via the Internet. So whenever an infrastructure fails, whatever the actual reason, the first finger often points at anonymous hackers taking down the system for fun and perhaps for profit. Sometimes these claims are even correct. But "sometimes," or even "occasionally," is too often for systems so critical to the functioning of modern industrialized society. Consequently, we have to understand how such cybershocks can happen and what we might do to minimize the damage they would cause.

As a starting point in "deconstructing" the problem of cybershock, it's useful to first understand how big the Internet really is in order to get a feel for what might be involved in totally shutting it down.

WHEN THE MUSIC STOPS

THE INTERNET IS ALMOST UNIMAGINABLY LARGE BY WHATEVER MEA-sure you care to employ. Here are a few statistical facts to ponder.

- As of mid-2008, there were over one trillion Internet addresses, far in excess of the world's population. (Note: This number is composed of Internet addresses, not just addresses on the World Wide Web, which has around two hundred million addresses—those beginning with www.) It would take more than thirty thousand years to even read all these addresses.
- There exist about 150 web addresses for each person alive on the planet today.
- The total information content of the Internet is about five million terabytes, or five billion gigabytes. To store this amount of data would require the capacity of about one million human brains. Put another way, this storage amounts to more than

one billion DVDs. To give another perspective on this vast number, in its decade or so of existence Google has managed to index approximately one-half of 1 percent of this data.

With these staggering statistics at hand, we see graphically the huge complexity of the Internet as a network of billions of nodes linked by many more billions of connections, all of which are dynamically coming and going every moment of each day.

COMEDIAN LOUIS CK HAS A STAND-UP ROUTINE ABOUT FLYING IN A plane equipped with high-speed Wi-Fi. Suddenly, the man sitting next to him breaks into an outburst against the airlines when he loses the connection. Louis CK asks, "How quickly does the world owe him something that he knew existed only ten seconds ago?" We humans indeed become accustomed very quickly to new technological gizmos and build them into our way of life almost overnight, particularly when they facilitate communication. Be it telephones, jet airplanes, or e-mail, we are hardwired for connecting with others—and the quicker, the better.

To find out how dependent today's man and woman are on the Internet, the chip maker Intel commissioned a survey of the matter a couple of years ago. The company asked more than two thousand men and women of all ages and walks of life whether they would rather forgo sex for two weeks or give up access to the Internet for the same period. Result? An amazing 46 percent of the women surveyed and 30 percent of the men opted to give up the sheets. Even more broadly, among all discretionary expenditure items—cable TV, eating out, fitness club membership, and even shopping for clothes (this one's really hard to believe)—the Internet ranked as the highest-priority item on the list. In total, almost two-thirds of the adults questioned said they simply could not live without the Internet.

Interestingly, in a similar survey by Dynamic Markets in 2003 of corporate employees and information technology managers in Europe and North America about the stress of being cut off from e-mail, it turned out that e-mail deprivation ranked as a higher stress inducer than . . . divorce. Or getting married. Or moving to a new residence. These people were then asked how long it would take after the e-mail went down before they would become angry. A full 20 percent said "Instantly!" And a whopping 82 percent of the group said they would be very angry by the end of one hour. In October 2010, Avanti Communications reported in a survey of companies worldwide that nearly 30 percent of the firms claimed they could not function without the Internet, while only a paltry 1 percent said that they could operate normally without it. The bottom line is clear: we not only love our Internet, but life as we know it literally cannot go on without it. Talk about a life-changing technology!

But web surfing and e-mails are conveniences, something that's usually not really a matter of life and death. How important is the Internet when it comes to more elemental existential matters like eating, drinking, earning a living, and staying healthy? Answer: Way beyond just *very* important; in fact, it borders on being life-threateningly important. To underscore this fact, here are but a few of the infrastructures we rely on every day that will vanish from our lives if the Internet goes down.

Personal and Commercial Financial Transactions: Whether you pay by credit card, check, or bank transfer, your money moves over the Internet. Of course, financial institutions have backups. But those require humans to process paperwork, which takes time, a lot of time compared with the rapidity needed to carry out a transaction via ATMs, e-banking, or Internet shopping.

At the "big money" level, the situation is far worse. While it's difficult to get a handle on the total volume of worldwide financial transactions processed daily through the Internet, a glimpse of the magnitude of these transactions is available by looking at the volume of daily foreign exchange trades. In 2007, the amount of money flowing through the system each day was nearly $4 trillion; by now, that amount must be pushing $10 trillion or more. And this is every trading day. What would happen if the Internet crashed and those transactions had to be done by fax, telephone, or even snail mail, as in the past? I shudder to think. One thing's for sure, and that is that life will be a mess worldwide for weeks, months, and possibly years after such a crash, even if the outages were only for a few days. Companies will fail, governments may collapse, and in general, chaos will reign supreme.

Retail Commerce: Almost all retail stores and supermarkets rely upon automated inventory control to keep the shelves stocked and ready for your shopping pleasure. For example, every time you buy an item at a chain store like H&M or at a bookshop like Barnes and Noble, the cash register immediately notifies a central computer of the item you purchased and the store's location, signaling the warehouse that a replacement item should be sent. That system—along with almost all retail commerce—would vanish within nanoseconds of the collapse of the Internet. The same applies to other retail outlets like gasoline stations, pharmacies, and food shops that provide the wherewithal for daily life.

To get a feel for the size of the problem, nearly $14 billion is spent each day in the USA alone in the course of a billion individual transactions—just for food and associated retail items. Only a minuscule fraction of these transactions could take place if the Internet communication system employed to log the transaction, update inventories, and the like were to fail.

Health Care: Almost all patient records are stored online. So doctors, hospitals, and pharmacies would have a difficult time accessing patient's records if the Internet went down. This, in turn, would lead to a major degradation in the immediate availability of health-care services. While you could probably still get health care in the absence of your records, what about without your health insurance card and/or the records standing behind it? Will the doctor or hospital welcome you when they have no way of verifying whether or not you can pay? No problem, you say. I'll pay in cash. Really? Where are you going to get that funds when the ATM machines are frozen and bank records inaccessible? You may have a tough time getting your hands on the kind of funds that health-care facilities will demand. So getting sick in an Internet-less environment will be far worse than it is today.

Transportation: Airlines and trains depend on the Internet for scheduling and monitoring their services. It's safe to say that a shutdown of the Internet would entail a shutdown of airports around the world, as well as huge scheduling problems for ground transport, including the trucks and trains that deliver the necessities of daily life to food stores and other retail shops.

This list could be considerably extended to encompass infrastructure failures of all types—communication, electric power, government services, corporate activity, and the like. But that would be overkill. This abbreviated list already makes the point transparently clear that every aspect of the lives we now take for granted would be seriously, and probably dramatically, imperiled by a major Internet failure. With these facts in mind, let's start looking at how such a failure might take place.

INTO THE HEART OF THE PROBLEM

POSSIBLE CRASHES OF THE INTERNET CAN BE CRUDELY SEPARATED into two categories: (1) *systemic crashes* due to the intrinsic limitations of the Internet structure itself and stress imposed by the exponentially increasing volume of traffic that the system is called upon to serve, and (2) *deliberate attacks* on the system by hackers, terror-ists, or other agencies intent on holding the Internet hostage to their wishes and goals. I'll address the second category in the next section. Into the first category fall both hardware and software failures. Here are a few not-so-well-chronicled examples that outline some of the possibilities.

Black Holes: When you can't reach a specific website at any given time, the usual reasons are that the site has been aban-doned, the server is down, the site is being maintained, or another easily explainable cause. But sometimes the site will simply not load. Occasionally, there does exist a path between your computer and the one hosting the site you're trying to reach, but the message gets lost along the way and falls into a "black hole" of information never to be seen again. Researchers have discovered that more than 7 percent of computers world-wide experienced this sort of error at least once during a three-week test in 2007. Some researchers estimate that more than two million of these temporary black holes come and go every day.

One of the reasons for these information sinks is routing difficulties stemming from the billion or so users of the In-ternet sending and receiving messages each day. As this traf-fic increases, routers responsible for matching up the message source with its intended recipient suffer a serious case of com-plexity overload, like a human brain that's called upon to pro-cess too many incoming requests and responses in too short a period of time. In the human case, ongoing stress of this sort may lead to a nervous breakdown. The Internet equivalent is a type of breakdown that some computer scientists, like Dmitri Krioukov from the University of California, San Diego, worry about, namely that the Internet in toto may collapse into a black hole.

This is a good juncture to mention another fine example

of a complexity mismatch. When the Internet was initially formed, people believed that the network (the links) would be dumb, but the end points (the nodes) would be smart. But maintaining security at the end points is proving to be a challenge, and we are beginning to see complexity overloads as each new type of security attack arrives. A Tainter-style collapse may well occur if people begin to lose faith in the Internet. They will not want to buy online, they will avoid social networking, and so forth. In essence, the Internet would fade into irrelevancy.

Power Consumption: The power consumed each day in carrying out the more than two billion Google searches adds up to electrical power usage in excess of what's consumed by three thousand households in Google's hometown of Mountain View, California. Now consider that YouTube, a Google subsidiary, accounts for over 10 percent of total Internet bandwidth. Add in social-networking sites like Facebook and Twitter, throw in video-streaming hosts such as Netflix, and you begin to get a feel for who are the bandwidth hogs of the Internet. Each of these services requires vast data centers, or "server farms," to handle this flood of bits and bytes that has to be moved through the network 24/7.

The heat produced by these data centers must be verified from the facilities housing the servers, in order to keep them at normal room temperatures of around 20 degrees (Celsius). This heat is generally just pumped outdoors rather than reused, making its own contribution to global warming. Moreover, the power consumed in cooling the data centers approximates the power consumption of the servers themselves. Ominously, this situation is growing by leaps and bounds, not declining. So if technological advances don't step in and put a halt to this "heat death," we could easily end up with data centers unable to cool themselves, in which case they would effectively melt down when the server CPUs or other hardware simply burns out.

The end result of that process is clear: when the data centers disappear, so does the Internet.

Cable Fragility: The optical fiber cables on the ocean floor that carry phone calls and Internet traffic around the world are less than one inch thick. This is a very thin line, literally and figuratively, upon which to base the foundation for a globally connected world. Interestingly, these cables break regularly. But there is generally no disruption of service when this happens, as the telecomm companies have backup systems in place and simply switch to alternative routes while the main lines are being repaired. But not always!

A good example of what can happen occurred in 2008 when two of the three cables passing through the Suez Canal were cut on the ocean floor near Alexandria, Egypt. This seriously disrupted phone and Internet service from the Middle East and India headed for Europe, forcing that traffic to take the eastward path around the globe instead.

Through accidents of geography and geopolitics, there are several choke points in the worldwide communication networks, Egypt being one of them. Since the cheapest way to carry the traffic over long distances is to put the cables underwater, a place like Egypt bordering on both the Mediterranean and the Red Sea, which in turn connects to the Indian Ocean, is an attractive choice. As a result, cables carrying information from Europe to India follow the route through the Suez Canal—just like ships. But Egypt is not the only choke point. The ocean floor off the coast of Taiwan is another, which accounts for why the December 2006 earthquake that severed seven of the eight cables in that region slowed down communication in Hong Kong and elsewhere in Asia for months until the cables could be repaired. Hawaii is still a third choke point for traffic connecting the United States to Australia and New Zealand. Any or all of these choke points form juicy targets of opportunity for slowing down the Internet in large areas of the world.

Router Scalability: Each minute, hundreds of Internet connection points drop offline. We don't notice since the network simply fences off the dropped connection links and creates a new route going around them. This reconfiguration takes place because the subnetworks making up the entire Internet communicate with one another through what are termed "routers." When a communication link changes, nearby routers inform their neighbors, who then transmit the knowledge to the entire network.

A few years ago, researchers in the United States came up with a method to interfere with the connection between two routers by disrupting the protocol they use to communicate, making it appear that the link between the routers is offline instead of active. Note that this disruption is local, disrupting just the connection link between a router and its immediate neighbors. Recently, though, Max Schuchard at the University of Minnesota and his colleagues discovered how to spread this disruption to the entire Internet.

Schuchard's technique is based on a denial-of-service (DOS)-style attack. This involves bombarding a particular website or sites with so much incoming traffic that the servers at the target site cannot handle the volume and shut down. Schuchard's experiment had a technical twist that would allow it to take down the entire Internet using a network of about a quarter of a million "slave" computers dedicated to the task. Details are beyond the scope of this book, but the general idea is to create more and more holes in the router network so that eventually communication becomes impossible. Schuchard says, "Once this attack got launched, it wouldn't be solved by technical means, but by network operators actually talking to each other." Restoration of Internet service would involve each subsystem being shut down and rebooted to clear the jam created by the DOS attack, a process that would take several days, if not longer. Is this procedure a viable way to take down the Internet?

An attacker who commands a quarter of a million "zombie" computers generally isn't going to use them to crash the Internet, but will instead employ that network for nefarious commercial purposes. But that rule doesn't apply to governments. One such scenario would be for a country to simply cut itself off from the Internet, like Egypt did during the uprising against the Mubarak regime in early 2011. That country could then launch an attack against an enemy, or for that matter, the remainder of the Internet, while keeping its own internal network in place.

In either case, Schuchard's work shows that regardless of who perpetrates such an attack there's not much that can be done at present to combat it. So far, nothing of this sort has come close to taking place. But, then, that's just what the study of X-events is about: surprising, damaging things that have yet to take place.

Router scalability serves as a lead-in to our second major Internet failure category, human error and/or malicious intent.

NOT BY ACCIDENT, BUT BY DESIGN

ONE MORNING IN APRIL 2009, MANY PEOPLE IN SILICON VALLEY woke up to discover that they no longer had phone, Internet, mobile, or cable television services. AT&T issued a public announcement that fiber-optic cables had been severed in many places. Speculation immediately began running rampant that these cuts had been perpetrated by workers who actually service the cables, since their union contract had expired just a few days before the service outage. Moreover, the cuts were sharp, not ragged, seeming to have been done with a hacksaw, and were within easy driving distance of each other. Paradoxically, what made the cuts even scarier was that they were easy to fix. It makes one wonder what might have happened to service if the perpetrators, whoever they were, had poured gasoline onto the cables and melted them. Or if a group of malcontents had formed together to coordinate such an attack to destroy fiber connections in areas where the density of cables is high.

All in all, this malicious hardware attack on the entire telecommunication infrastructure, including sawing through the Internet cables, gave new meaning to the term "hacker." Of course, what's usually meant when this expression comes up is an attack based on meddling with software on the Internet, not physically destroying underground or underwater hardware installations. So let's have a quick tour of a few common ways to kill the Internet softly by disinformation and destructive program alterations rather than by the destruction of matter.

Certainly the most well-publicized type of software attack is a virus of some sort. Just like their biological namesakes, computer viruses take over the operating system of their host computers and force them to carry out the instructions coded into the virus rather than the instructions from the machine's own operating system. In late 2009, a vicious little devil called the Stuxnet virus infected forty-five thousand computers worldwide, displaying a pronounced fondness for industrial control machines built by Siemens AG of Germany that were being used mostly in Iran. Since the Iranians were using this equipment as part of their nuclear power program (and probably nuclear weapons development, as well), most knowledgeable people felt the attack was created by individuals working for a country or a well-financed private organization focusing on the disruption of Iranian nuclear research. (For aficionados, the Stuxnet was what is technically termed a "worm" rather than a virus. But for our purposes it makes no difference.)

Stuxnet was a malicious piece of computer code that heralded a new form of warfare: killing by false information instead of by guns and bombs. Why send troops to knock out critical infrastructure (i.e., power plants or water-treatment facilities) when you can do it remotely from halfway around the world by bits and bytes? As more and more military operations are conducted virtually by drones like the US Predator aircraft, it's possible such weapons could be compromised and used on friendly troops instead. This is to say nothing of the conceivable breaches of national security systems and intelligence networks. Or, for that matter, the command and control of nuclear weapons, about which I'll say more later.

Also, we cannot rule out the existence of another Kaminsky-style glitch lurking somewhere in the deep structure of the Internet, something different from the DNS problem he uncovered but equally dangerous. Of course, this sort of "Kaminskyesque" flaw falls into the category of an "unknown unknown," vaguely analogous to the threat of an alien invasion. Like dark invaders from outer space, a design flaw deep in the Internet may or may not exist. And even if it does, it may never turn up on our doorstep. There is simply no way to rationally evaluate this possibility. So for now I file it with the other unknown unknowns and move on to a second way of disabling parts of the Internet: a far-reaching denial-of-service (DOS) attack.

ON JULY 4, 2009, SEVERAL US GOVERNMENT AGENCY COMPUTERS were subject to massive DOS attacks that went on for several days. Systems affected included those at the Treasury Department, the Secret Service, the Federal Trade Commission, and the Department of Transportation. According to private website monitoring firms, the DOT website was 100 percent down for two full days, so that no users could get through to it during one of the country's heaviest travel weekends. According to Ben Rushlo, director of Internet technologies at Keynote Systems, a firm that keeps track of website outages, "This is very strange. You don't see this. Having something 100 percent down for a 24-hour-plus period is a pretty significant event. The fact that it lasted for so long and that it was so significant in its ability to bring the site down says something about the site's ability to fend off [an attack] or about the severity of the attack."

In fact, DOS attacks are not at all uncommon, even though it's notoriously difficult to measure just how many such attacks regularly occur. In 2005, Jelena Mirkovic and her colleagues estimated the number at about twelve thousand per week. And surely this level has not decreased since that time. Moreover, DOS attacks are relatively easy to mount using widely available hacking programs. They can be made even more damaging if thousands of computers are coordinated for the attack, so that each of the computers is sending messages to the target. This is the very sort of attack I mentioned earlier that knocked out computer systems in Estonia. A similar assault took place in Georgia during the weeks leading up to the war between Russia and Georgia, when the Georgian government and corporate sites experienced outages that were again attributed to attacks by the Russian government. The Kremlin, of course, denied responsibility. But independent Western experts traced the incoming traffic to specific domain names and web registration data, concluding that the Russian security and military agencies were indeed the perpetrators of this particular attack.

To bring our understanding of DOS attacks down to daily life, the social networking service Twitter was totally knocked out for several hours in 2009 by such an attack carried out by a lone blogger, coincidentally also located in the Republic of Georgia. The attack was targeted at a blogger whose handle is "Cyxymu," Cyrillic spelling for the town of Sukhumi in the breakaway territory of Abkhazia. According to Ray Dickenson, chief technology officer at Authentium, a computer security firm, "It's as if a viewer who didn't like one show on a television channel decided to knock out the whole station."

Viruses/worms and DOS attacks are headline grabbers, probably because they target the Internet at the level at which users actually interface with the system—their own computers and/or the service providers. Attacks at this level are good for exciting the media and are easy for just about any Internet user to understand and relate to in his or her daily life. But while not impossible, it's unlikely that the Internet as a whole is threatened by such "surface" attacks. To bring down the entire Internet, or even a major chunk of it, you have to dig a lot deeper into the system, as described in the story I told earlier about

Dan Kaminsky and the DNS hole in the network. Or perhaps you have to assemble a global team of hackers of the type that broke into the networks of Citibank, the IRS, PBS television, and other major financial and media organization following the WikiLeaks events in 2011.

ADDING IT ALL UP

IN 2006, COMPUTER SECURITY EXPERT NOAM EPPEL PUBLISHED AN article on the Internet titled "Security Absurdity: The Complete, Unquestionable, and Total Failure of Information Security." As one might imagine from the title, this piece attracted a lot of attention from professionals and firms dealing with Internet security. (As an amusing footnote to underscore the problems Eppel identified, during the writing of this chapter I tried finding that original article to have a look at recent comments that may have been added since I downloaded the piece in late 2007. To my astonishment, I discovered that every Google hit in my search for the article sent me to a website called www.securityabsurdity.com, which appears to be a site that's hijacked whatever the original site was that contained the actual article. Needless to say, Eppel's article was nowhere to be found.)

Eppel identifies sixteen different categories of security failures that infest the Internet. Included among the items on his hit list are spyware, viruses/worms, spam, and DOS attacks. As far as I can tell, very little, if anything, has been done to address effectively any of the problems arising from even one of the sixteen categories on Eppel's list. As he noted, the situation is very much like the story of the frog in a pot of boiling water. If you place the frog in the pot when the water is cold and then gradually bring it to boiling, the frog sinks into a torpid state after a bit as the water heats up, and ultimately sits quietly as it boils to death. According to Eppel's story, the computer security industry is analogous to that frog. The system is dying. But the death is tolerated simply because we're accustomed to it. In short, the security industry is failing in every possible way because it is being outinnovated. And who is doing that innovating? Answer: A vast community of suppliers of so-called security systems, computer criminals, spammers, and others of this ilk, not to mention the willing complicity of computer users who buy in to the flimflam peddled by the "professionals."

Just to get a feel for the severity of the problem of Internet security for the everyday user, studies have been carried out to determine the time it takes for a brand-new "virgin" computer to become infected with some sort of spyware, virus, identity theft, or other sort of malware from the moment it's plugged in, booted up, and connected to the Internet. The average time to infection turns out to be about four minutes! It's reported that in some cases the time before someone else takes complete control of the computer turning it into a "zombie" is as short as thirty seconds! There is little doubt that what we're facing is not a security epidemic, but a full-fledged pandemic.

Even in the face of these sorts of results (and you can do the experiment yourself if you don't believe them), a quick look at sites that report Internet security breaches in real time will show there's nothing amiss. For instance, I've just finished looking at the real-time threat-monitoring sites from firms selling antivirus packages (to keep the lawyers away, I'll refrain from mentioning names so as to protect the guilty). Looking at their threat maps for security problems around the world, you'll see the odd blip here and there. But in each case, the overall threat level for the Internet as a whole is reported in at most the yellow zone, and for most regions it's solidly in the green. Yet on a casual web surf using the term "Internet security threat," I turned up article after article saying that the number of threats is dramatically increasing from the previous year's level. What's amusing, if not worrying, is that some of these articles have been prepared by the very same firms whose threat maps never show the Internet under assault. If this isn't a living example of a frog in the pot of water, I don't know what is.

It's worth noting that the needs and wishes of the computer security business are not the only element acting to lock us into the existing Internet. Technology companies, too, are trapped. They have to sell today's products and there is high uncertainty in making investments in new technology. Corporate information administrators have to defend past purchasing decisions. So how do we carry out a "makeover," or introduce an entirely new Internet? The US National Science Foundation's GENI Project has created a virtual laboratory for exploring future Internets at full scale and aims to create major opportunities for understanding, innovating, and transforming global networks and their interactions with society. Other private groups are exploring the same territory, with the goal of figuring out how to smoothly transit from the existing Internet to a much safer, more user-friendly version without throwing out the baby with the bathwater.

The bottom line here is that there is no such thing as true security on the Internet. At some degree of everyday usage, the Internet functions without any obvious holes. But that doesn't mean the holes aren't there. And it doesn't mean they're not getting bigger. The question is when will they get so big that too many people, corporations, or governments fall into them and can't crawl back out. When that day comes, the Internet's days are numbered, at least the Internet as we know it today. The current system is using a 1970s architecture to try to serve twenty-first-century needs that were never envisioned in those halcyon days of a bipolar world. (Try using a 1970s computer today to access the modern Internet!) The two systems in interaction have created a huge complexity gap that's widening by the day. Soon it will have to be narrowed—by hook or by crash.

WHEN DO WE EAT?

BREAKDOWN OF THE GLOBAL FOOD-SUPPLY SYSTEM

FOOD FACTS
DID YOU KNOW THE FOLLOWING?

- Over forty-four million people have been driven to poverty since June 2011 due to the rising price of food.
- Due to decreasing water supplies, Saudi Arabia will no longer be able to produce wheat after 2012.
- Since the way we produce and transport food today is heavily dependent on oil, food prices will continue to track the price of oil, and if that price stays high, some forms of food production will no longer be economically feasible.
- According to studies by the Global Phosphorus Research Initiative, within the next two or three decades there will not be enough phosphorus to meet food production requirements.
- The world price of food has risen nearly 40 percent since early 2010.

- Diseases like UG99 wheat rust are destroying ever larger segments of the world's food supply.

Well, I could go on like this for several more pages. But there is little doubt that the world's food-supply chain is threatened by massive "complexification" of the food industry. Industrialization of farming, genetic modifications, pesticides, monoculture, climate instability, population growth, encroachment of cities on farm land (and their water)—take your pick. All these are individually and collectively creating conditions for an X-event-type collapse of the world's food production and distribution network. Industrialization of food has led to an overreliance on certain crops—corn, wheat, soybeans—leaving us with a dramatically reduced natural diversity, a diversity that has long shielded us from plagues, climatic variation, and the like. Some fear that pesticides are contributing to the evolution of "superpests" (just as antibiotics led to superviruses) that could ravage the globe, immune to all attempts to stop them.

The competing complexities in the different parts of the global food system seem much more likely to increase the complexity overload on the system rather than reduce it unless more, much more, international cooperation takes place to reduce the imbalances. Otherwise, the third horseman of the apocalypse, famine, will ride roughshod over the globe and force the world into facing the food problem on far less favorable terms than we enjoy today.

Most of us living in the industrialized world are accustomed to supermarket shelves sagging under the weight of inexpensive food of a bewildering variety of types. It's hard for us to imagine that life could ever be different. But like a lot of other eras I cover in this book, the era of easily accessible, cheap food is one that is ending, as we embark on a journey to a time when there simply won't be enough food for everyone. If you're starting to think this is just another Malthusian scare story waiting to be quashed by another "green revolution," read on.

BLACK PLAGUE 2.0

IN THE LATE 1800S, A FUNGUS ORIGINALLY FROM THE HIMALAYAS traveled to Europe from the Dutch East Indies and later moved to North America, where it ran through the forests of eastern Canada and the United States killing the vast majority of elm trees, earning the sobriquet Dutch elm disease from its identification by Dutch scientists in 1917. New strains emerged in the 1970s in the United Kingdom, where over three-quarters of the elms perished. Now meet its much more aggressive younger brother, *Phytophthora ramorum* (PR), another fungus pathogen also thought to have started in Asia and moved to Europe and the United Kingdom via shipping containers in the 1990s. But instead of the elms, which have pretty much disappeared, PR is attacking the larch trees that cover rural regions in Devon, Cornwall, and South Wales.

As with some human afflictions, such as ovarian cancer, the first you know about a PR infection is when it's already too late to save the tree. The outward signs begin when the tree's foliage starts to blacken, the inner bark turns brown instead of green, and a black liquid oozes out of the tree from various breaches in its bark. At this point, the tree cannot be saved and must be cut down and removed in an effort to protect the rest of the forest. And it's not just larch trees that can be infected. The PR fungus can also attack beech, sweet and horse chestnut trees, as well as a range of plants that include rhododendron, lilac, and viburnum.

The first appearance of PR in the United Kingdom was in 2002 on a viburnum plant in East Sussex. The fungus then "leapt" to rhododendrons, and from there to other species of plants through the rhododendron spores that traveled in water, air, and mists to other plants. Until 2009 tree scientists saw only about a hundred infected trees, usually near rhododendrons. But then PR spores began sprouting up all over Wales, Northern Ireland, and the Irish Republic. Even worse was that the spores in the trees were reproducing more than five times faster than on rhododendrons. So the race to save the forests was on, pitting the skills of the scientists against the virulence of the PR spores.

If you stand on a hilltop overlooking a forest in South Wales today, what you see is more reminiscent of a World War I battlefield than a forest, as the acres of stumps and branches of cut trees give testament to the fact that there is still no known cure for the PR fungus other than the "cut-and-burn," scorched-earth strategy of ancient times. By February 2011, nearly 1.5 million larches had been cut in the preceding fifteen months, with another 1.2 million due for the chopping block over the following three months—in order to stave off an even worse catastrophe.

What's even more disturbing is that the PR pathogen remains in the soil for at least five years. So the future of larch trees in the United Kingdom looks distinctly bleak. And the big fear is that once the larches are removed, PR will jump to yet another species. Some informed analysts suggest that blueberries or even heather are strong possibilities. Currently, plant experts seem to have abandoned the idea of actually eradicating PR and are focusing their efforts on simply containing it. Truly, a plague of biblical proportions is brewing in the woodlands of Britain, but one unlike the plague of the Middle Ages in that it's unlikely to disappear anytime soon. To get a sense for what the world might be like if a plant killer like PR started running amok over the entire globe, consider the following story.

SOME YEARS BEFORE ONLINE BOOKSELLERS LIKE AMAZON MADE tracking down just about any book in or out of print as easy as a few clicks of a mouse, I undertook a worldwide odyssey through second-hand bookshops from New York to Christchurch to Rio de Janeiro in an effort to find the one hundred greatest science-fiction novels of all time. This quixotic undertaking was sparked by the 1986 book *Science Fiction: The 100 Best Novels,* by sci-fi critic and publisher David Pringle. To Pringle's credit, he didn't try to rank-order his top one hundred, but simply listed them by order of publication, starting with *Nineteen Eight-Four* by George Orwell (1949) and ending with William Gibson's *Neuromancer,* which coincidentally was published in . . . 1984. Somewhere in between these marvelous classics, both of which trace out a somewhat dark vision of the future, was the equally classic—and equally dark—1956 tale *The Death of Grass* by British writer John Christopher.

The premise of Christopher's story is that an uncontrollable plant virus, the Chung-Li, wipes out all grasses in China, leading to the starvation of hundreds of millions of Chinese. This all seems hopelessly remote to the middle-class Custance family in Britain, one branch of which tills the soil in the Lake District; the other branch, headed by lawyer John Custance, lives and plays in London. As the story unfolds, John Custance has heard from Roger, a friend in the British government, that the Chung-Li virus has spread to Britain. The following exchange captures the essence of the story:

> *"Damn it!" John said. "This isn't China."*
> *"No," Roger said. "This is a country of fifty million people that imports near half its food requirements."*
> *"We might have to tighten our belts."*
> *"A tight belt," said Roger, "looks silly on a skeleton."*

All scientific attempts to stem the virus come to nothing, and after a year the entire world is affected. John then hears from his friend in government that the army is about to seal off the major cities, since only a small fraction of the population can survive on root plants and fish. So the government has decided that the only solution is to "downsize" the population by killing off all city dwellers. Most of the book is then devoted to John's journey with his family to his brother's farm. On the way, they encounter marauding bands of city folks like themselves, who have escaped the cities and are raping, looting, and murdering their way through the countryside. John's own party has to rob and kill in order to survive. Ultimately, John and his family

reach the haven of his brother's farm, which is located in a secluded—
and protected—valley where they feel they'll be better positioned to
defend their turf against all comers.

As a real-world example of an endgame situation, albeit
one not quite as apocalyptic as that portrayed in *The Death of Grass*,
the PR fungus discussed earlier could possibly drive larch trees,
rhododendrons, and their like to extinction. Although the "death
of larch trees" is hardly an existential threat, such a fungus could
spread out to a global catastrophe if it were to mutate and threaten
grain crops as well. In recognition of this general possibility of plant
species extinction, in 2008 the government of Norway established a
"doomsday" seed vault deep inside a mountain in the Arctic archi-
pelago of Svalbard, about one thousand kilometers from the North
Pole. As Norway's prime minister Jens Stoltenberg stated, "It is the
Noah's Ark for securing biological diversity for future generations."
Buried within the permafrost of a mountain, the vault is designed
to withstand earthquakes, a nuclear strike, perhaps even an asteroid
impact.

The underlying motivation for such a vault rests with the industrialization of the global food supply. Big firms that dominate food
production severely restrict genetic diversity by employing just a
few varieties of plant seed, sometimes just a *single* variety. If a disease struck that particular variety, food production would be in big
trouble, and the entire food-supply system could collapse. Hence, the
Svalbard vault.

Even though the Arctic region would seem to be plenty cold
enough to preserve the seeds, the temperatures there apparently vary
sufficiently widely that the doomsday vault needs huge air conditioners to chill the vault to temperatures just below zero degrees Fahrenheit, in order to deep-freeze the seeds to a temperature where they can
survive for a thousand years. The seeds themselves are packed in sil-

very foil containers and placed on blue and orange metal shelves inside the storage rooms. A total of about four and a half million seed types ranging from carrots to wheat to corn can be stored. For the record, the first seeds deposited at the opening ceremony were a collection of rice seeds from 104 countries.

To most of us, I think, it will come as a surprise to know that there already exist about fourteen hundred seed banks around the world. But many of them are in politically unstable areas or face threats from the natural environment where they're located. For example, seed banks in Iraq and Afghanistan were wiped out by war, and one in the Philippines was destroyed by a typhoon in 2006. Thus, it was important to establish a kind of "last resort" seed bank that could withstand just about everything nature and humans could throw at it. As stated by Geoff Hawtin of the Global Crop Diversity Trust, who organized and funded the operation, "What will go into the cave is a copy of all the material that is currently in collections [spread] all around the world." Even though Norway is the vault's formal owner, any country may deposit seeds without charge into the doomsday vault and reserve the right to withdraw them as needed.

A plant virus like the Chung-Li strikes right at the deepest level of the food chain beginning with grasses, destroying that vital first link. But a virus isn't the only way to undo the plant world as there are other links in the chain from grass to the foodstuffs that end on your dinner plate. Here's one of those links that's also of deep concern today.

THE DEATH OF BEES

ONE OF THE MOST ACCLAIMED DOCUMENTARY FILMS IN RECENT YEARS was *The Vanishing of the Bees,* which told the tale of the mystifying disappearance of over one-third of the honeybees in North America and Europe from 2006 through 2008. This vanishing act, termed "colony collapse disorder" (CCD), is portrayed in the film by following actual beekeepers, as we observe them opening their hives in the morning and seeing that all the bees have flown the coop, so to speak, literally overnight.

The idea that bees are disappearing has now embedded itself into the public consciousness, and the film plays into that fear in many ways. There's no denying that pollination of plants by bees, as well as other animals like butterflies and birds, plays a crucial role for the production of fruits and seeds. Over 80 percent of the quarter of a million flowering plants on the planet are pollinated by such animals. It is also undeniable that honeybees, which are the main pollinators among the many species of bees, suffered a huge die-off beginning in 2006. In effect, disappearance of the honeybees constitutes a huge reduction in the complexity of the food production process, creating a complexity imbalance between the variety of tools needed to pollinate plants and those available, since without the honeybees the process would have to be carried out by a smaller set of pollinators.

These facts raise two huge questions for the human food-supply system: (1) Why are the honeybees dying? and (2) How crucial is bee pollination in the overall scheme of things as far as food production is concerned? Well, let's see.

The honeybees have a hard life in today's world. They are trucked around the country by their keepers to pollinate fruit and nut crops, starting with almonds in California in early spring and ending with pears and apples in Oregon in the early autumn. Note here that these are the so-called commercial honeybees, which far outdistance bees from the wild, the feral bees, as pollinators. To get an idea of the damage that would be caused if these bees didn't show up any longer, consider that California provides nearly 80 percent of the world's almond supplies, which are used in products ranging from ice cream to cosmetics. So it's no surprise to hear that firms like Häagen-Dazs are funding efforts to raise public awareness of the importance of bees in agriculture. To put the matter bluntly, no bees, no crops, no products.

To get an idea of how CCD has impacted the economics of honeybees, almond growers paid a rental fee of about $175 per hive in 2009, which was more than double the price just four years earlier. So if you place one hive per acre of planted trees and you have 2,000 acres, then you're looking at an increase in production costs of $200,000 or more in pollination fees alone. As industry professional John Replogle, former CEO of Burt's Bees, a cosmetics firm that offers almond-based creams, put it: "So go the bees, so go the almonds." And this economic vignette is just for almonds. Scale it up to apples, pears, blueberries, and zillions of other fruits, nuts, and plants that bees pollinate and you begin to get some sense of the magnitude of the problem if the bees go AWOL. What do conservation biologists and bee specialists say about the causes of CCD?

Following the CCD outbreak in 2006, researchers worked overtime to try to identify just what it is that caused the bees to flee. In early 2011, the betting odds seemed to favor an explanation that it's in the genes; the factories in the cell that create the proteins bees use to carry on their activities seem to have broken down in those bees associated with CCD. To put it succinctly, the bees' cellular structure couldn't manufacture the energy necessary for the bees to function. But what were the factors that gave rise to this breakdown in the cellular energy-manufacturing facilities in the first place?

The best account going is that three rather different factors combined in a perfect storm of sorts to interfere with the affected bees' genetic operation. This infernal triangle consists of the following steps.

- *Pesticides*: The downside of pesticides has been well chronicled since at least 1962, the year of publication of Rachel Carson's eye-opening book *Silent Spring,* which drew attention to the dangers for both humans and the environment of chemical pesticides. Like all blunt tools, pesticides have two faces. They can kill mosquitoes that breed deadly diseases. But they can also kill ants and other helpful insects. They can prevent illness in humans by destroying diseased foods. But they can be carcinogenic as well, leading to threats like breast cancer. So their use is always a double-edged sword.

In this same direction, genetic modification of crop seed to reduce the need for insecticides have given rise to an evolutionary arms race between the farmers and the insects, whose latest manifestation is a population of "superinsects" that can resist the genetically implanted pesticide placed into popular strains of corn. Yet once again this is an example of complexity overload, in which the immune system of the insects has evolved to a level of complexity far greater than the low-complexity level of the genetically engineered defenses hardwired in to the plant's genome.

Traveling bees, too, are exposed to increasing levels of pesticides as each year goes by. This exposure, coupled with the simple stress of being pollinating traveling salesmen, takes a toll on the health of the bees, reducing their ability to resist other pathogens.

- *Viruses*: Several viruses are known to be damaging to bees. The list of viruses that attack the genetic structure of bees is a long one, including something called the Israeli paralysis virus and a parasitic fungus called *Nosema ceranae,* both of which regularly appear in the genetic makeup of infected bees. Here, again, complexity rears its head; as the variety of threatening viruses multiplies, the complexity of the bees' immune system is outmatched by the variety in the threats leading to a gap that will ultimately have to be narrowed.

- *Varroa mites*: The tipping factor that may well have precipitated the colony collapse by bees weakened from the foregoing factors is the varroa mite that was accidentally introduced into the United States in 1986, almost surely through importation of infected bees. This mite is a known carrier of the very type of viruses that the genes of bees are particularly susceptible to, and it may well have been the straw that broke the bees' back leading to CCD.

Fortunately, bee populations seem to be on the rise again. But new pollination problems are looming on the horizon due to the dramatic increase in total agricultural production over the past five decades. In that time, human population has doubled. But the small proportion of agricultural output that depends on bee pollination has quadrupled over the same period. This heightened agricultural output of crops like cashews, cherries, and almonds has happened mostly by increasing the amount of land under cultivation.

Unfortunately, creating cultivated land from the natural habitat of wild pollinators, along with increased demand for pollinators, dramatically outstrips the increase in population of the domestic bees, which in turn is putting a serious strain on pollination capacity. So the rise in demand for pollinator-dependent crops, coupled with these factors reducing the pollinating capacity, has created the potential for a problem of unprecedented magnitude. The good news is that the bees are back; the bad news is that recent events may be an early-warning signal that we will have a real pollination problem on our hands very soon.

The PR fungus and the disappearance of bees vividly illustrate the need for a global plant seed backup like the doomsday vault to preserve the diversity of plant life. But food crises don't just come from problems with plants; they can arise in many ways and on many timescales. Reduced diversity from disease and infection is one of the more dramatic types of threats, generally unfolding on a medium-term timescale of a few months to a year or so. But one might well argue that these are at best borderline examples of how food-supply crises might arise from human actions. So let's take a harder look at the emerging food crisis, one whose focus is more short term, and see what kinds of storms appear to be brewing up on the horizon.

FAM[IN]E AND [MIS]FORTUNE

Shortly after ten o'clock on the morning of April 24, 2008, Mary Ann Galviso, a real estate broker from the rural community of Orosi in central California, snapped up two fifty-pound bags of

Thai jasmine rice from the San Francisco branch of the warehouse club Costco. Her purchase was a small contribution to the emptying of three full pallets of rice at the club within an hour's time, despite the fact that the store had imposed a limit of two bags per customer. What made Ms. Galviso's story marginally noteworthy was that she had traveled more than two hundred miles from her home to make this purchase, as the local store in Orosi had run out of the rice that served as a staple in meals for Mary Ann and her family.

This story highlights hoarding behavior on the part of not only home shoppers, but Asian and Indian restaurant owners as well, whose panic buying forced not only Costco, but also Sam's Club, a branch of Walmart, to impose limits on the amount of rice that customers could buy. One manager at Costco stated, "We've heard of cases where restaurant owners are hoarding three weeks' supply of rice in their basement."

Rice rationing in the United States in 2008 came as commodity price increases sparked violence over food supplies and costs. And, in fact, three years later the price increases had not abated and contributed mightily to the social unrest and violence that led to the ouster of the political regimes in Tunisia and Egypt. To appreciate the magnitude of the problem, in February 2011, the Food and Agriculture Organization of the United Nations reported that its food price index composed of a basket of key commodities such as wheat, milk, oil, and sugar had risen 2.2 percent from its January level and sat at its highest level since the organization started monitoring prices more than two decades earlier. Let's take a quick look at the constellation of problems leading to this continuing escalation of food prices worldwide.

The huge increase in food prices over the past five years can be accounted for by factors acting to simultaneously reduce the supply of food and greatly increase its demand. As we all know from Economics 101, these two factors constitute the perfect one-two combo to punch a hole in the food budget of every household. The supply-side factors leading to a decline in food production include the following:

Water Shortages: Overpumping of underground water aquifers in many countries, including China, India, and the United States, has artificially inflated food production in the past few decades. For example, Saudi Arabia was self-sufficient in wheat harvesting for over twenty years. Now the wheat harvest there is likely to disappear entirely over the next couple of years due to a lack of underground water to irrigate crops.

Soil Erosion and Loss of Croplands: Experts estimate that a third of the world's cropland is losing topsoil faster than it can be replaced through natural processes. In northwestern China and western Mongolia, a huge dust bowl is forming that will make the dust bowls in the United States during the Great Depression look minuscule by comparison. A similar dust bowl is also growing in central Africa. So in these regions, grain harvests are shrinking and farmers will eventually have to abandon the land and move to the cities.

Extreme Weather and Climate Phenomena: Global warming is not a myth, and rising temperatures are here to stay. It's estimated that for each one degree Celsius rise in temperature above the optimum for the growing season, there is a 10 percent decline in the yield of grain. Photos of burnt-out wheat fields in Russia during the summer of 2010 give graphic evidence of the impact climate change is making in dramatically reducing harvests.

High Oil Prices: There is a second "oil shock" taking place alongside the one you see at the gas station when you fill up your tank. This is the dramatic rise in the price of cooking oils like palm oil, soybean oil, corn oil, and many other vegetable oils. And there is also the strong impact a rising price of petroleum products makes on food supply, since oil enters into every aspect of food supply. As one wag put it, "Soil is nature's way of transforming oil into food."

These are but a handful of factors, all contributing to decline in global food supply. Unfortunately, there is an equally "depressive" complementary list on the demand side of the ledger.

Population Growth: The world population is growing at a rate of over two hundred thousand people per day. It doesn't take an undiscovered genius to know that most of them are entering the dining room in the poorest, most undernourished parts of the world. So while the number of mouths to feed continues to grow, the food available just isn't there to fill them.

Rising Affluence: Over three billion people are changing their dietary preferences from subsistence diets to more meat, eggs, milk, and grain-intensive products. For instance, it takes up to sixteen pounds of grain to produce just one pound of meat. And more than eleven times as much fossil fuel is needed to make one calorie of animal protein than what's required to make one calorie of plant protein. In other words, instead of eating the grains directly, half the world's population is eating "transformed" grain in the form of higher protein, but higher grain content, products like meat and poultry.

Nobel Prize–winning economist Amartya Sen illustrates the perversity of this phenomenon by envisioning a country with a lot of poor people that suddenly experiences economic expansion. He then assumes that only half the population shares in this newfound wealth. The rich half spend their wealth on higher-quality, higher-priced food, which drives food prices up. The poor half of the population now faces higher prices but without higher income and thus starves. And this is no fantasy either, as Sen traces just this process as having unfolded in the Bengal famine of 1943.

Grains to Fuel: A substantial fraction of grain, especially corn, is now being diverted from food to the production of ethanol to fuel cars. In the United States, this amounts to nearly one-third of the entire grain harvest. The primary motivation for this diversion rests with misguided policies instituted in 2006 that offered government subsidies to farmers who would convert their grain to automotive fuel.

In addition to these components of food price increases, we have the massive infusion of US dollars into the world's financial system to combat the ongoing financial crisis. Since commodities are almost all priced in dollars, dollars flooding into the world's financial system contribute to a huge increase in the price of commodity products worldwide ranging from oil to wheat to frozen pork bellies. As the value of the dollar rises against other currencies, holders of those currencies must pay more for their commodities and thus more for their basket of food. In short, the quantitative "easing" imposed by the US Federal Reserve has become a quantitative "hardship" at the dinner table for much of the world. So what is the solution? Or is there one?

The obvious "solution" to the food crisis, such as it is, would be to attack the supply and demand imbalances at their source. This would involve the following types of actions:

Efficient Management of Water Resources and Cropland: Overpumping of water aquifers and overstressing croplands must both stop. Suburban sprawl and paving over cropland for roads and parking lots, especially in China and India, has to be curbed. Furthermore, the water that is available must be employed in a much more efficient manner, which will require innovative technologies, technologies that currently do not exist: for instance, recycling of water, or the use of less-water-intensive crops.

World Population: There is a need to dramatically speed up the trend to smaller families. Among other things, this will require providing family-planning help and advice to the world's poorest populations, which have the highest birthrates.

Climate: Carbon emissions need to be cut by 80 percent over the next decade, in order to stave off catastrophic climate events, especially the flooding, hurricanes, droughts, and similar events that seem likely to arise from increased global warming. These events would destroy much current agricultural land.

Government Policy Change: Misguided government policies subsidizing ethanol production must be dropped as soon as possible. Those grain harvests are much more valuable filling empty stomachs than empty gas tanks. Increased governmental regulation of multinational food producers would also be a boon toward stabilizing food production by encouraging less efficiency and more resiliency in the production process, as for instance by forcing the usage of a greater variety of seed types in the production of grain.

ADDING IT ALL UP

THE WORLD IS CURRENTLY FACING A CONFLUENCE OF MOUNTING shortages in three commodities essential to the continuance of human life on this planet: water, energy, and food. These three elements combine into something much greater than the sum of its parts, a looming global disaster by 2030. By 2030 the demand for water will increase by 30 percent, while demands for both energy and food will shoot up 50 percent. All this will be driven by a global population increase to about 8 billion people, placing tremendous stress on our highly industrialized global food system.

It's important to note that the food shortage leg of this triangle is not being driven by supply-side considerations as much as it is by the demand side. Luckily, even though increasing population is a major factor in stimulating this demand, the rate of world population growth appears to be slowing, and if current patterns of women's empowerment continue, it can be expected to slow even further. Of course, surprises always occur so there is no guarantee that either of these trends will continue. And, in fact, they almost surely will not continue if the world fails to face up to the most challenging problem, which is the ever-increasing gap between the rich countries and the poor ones in the global economy.

THE DAY THE ELECTRONICS DIED

A CONTINENT-WIDE ELECTROMAGNETIC PULSE DESTROYS ALL ELECTRONICS

LESS THAN A MILLISECOND

GERMAN FILM DIRECTOR WIM WENDERS IS KNOWN FOR HIS "ROAD" movies, in which characters wander through desolate landscapes grappling with various sorts of existential angst. His 1991 film *Until the End of the World* is set at the turn of the millennium and involves an out-of-control nuclear-powered satellite that's about to reenter the atmosphere at some unknown location and contaminate a large area of the planet. People in different possible impact locations start panicking and flee in large numbers. Amid a lot of running around in the desert by the film's heroine, a woman named Claire, there's a collection of mad scientists, shadowy characters from covert government agencies, hitchhikers, bounty hunters, and other assorted riffraff looking to recover the prototype of a device for recording and translating brain impulses. During the course of this manic chase, the nuclear satellite is shot down, causing a huge burst of energy—an electromagnetic pulse (EMP)—that wipes out all unshielded electronic equipment throughout the world. As a result, the characters are transported from the late twentieth century back to a kind of caveman style of existence, as all devices worldwide that rely on microcircuits, such as computers, cars, radios, and the like are burned out in the span of a few milliseconds.

Is this something that could really happen? Could all the electronics we take for granted in everyday life suddenly be shorted out? Or is an EMP, Wenders-style, just another wild Hollywood extrapolation of something that's theoretically possible to something we should all be concerned about? A bit of history will quickly sort out the matter.

On July 16, 1997, Representative Curt Weldon, chairman of the House Military Research and Development Subcommittee, convened a session on "The Threat Posed by Electromagnetic Pulse (EMP) to U.S. Military Systems and Civil Infrastructure." Among the experts called upon for testimony were Dr. Lowell Wood of the Livermore National Laboratory in California; Gilbert Clinger, acting deputy undersecretary of defense for space; and Dr. Gary Smith, director of the Applied Physics Laboratory at Johns Hopkins University. Others testifying included several members from the US intelligence community. The conclusion of the hearing can be captured by a statement from Dr. Wood near the end of the session:

> It is a reasonable projection that most, if not all, modern computer systems exposed to EMP field levels . . . would wilt. By wilting, they would at least cease to function. In many cases they would be burned out . . . Not just computers in aircraft but computers everywhere, other than this type of very high integrity metallic enclosures that Dr. Ullrich sketched in his opening statement. Computers in any other enclosure than that type would be compromised, if not destroyed outright.

Later, Congress charged a blue-ribbon committee with undertaking a deeper investigation of the entire EMP phenomenon, which was published in 2004 as *Report of the Commission to Assess the Threat to the United States from Electromagnetic Pulse (EMP) Attack.*

In view of these detailed studies, we can safely conclude that not only is the EMP a viable threat to a modern, high-tech way of life, it is only going to become more threatening as we become ever more dependent on increasingly delicate electronics to navigate through our daily lives.

WHAT IS AN EMP, ANYWAY?

To put it very compactly, an EMP is an electromagnetic shock wave produced by a very high-energy explosion in the atmosphere. This wave creates a momentary surge of electrical current in the circuitry of any device like a cell phone, computer, television set, or automobile that is not shielded for protection. This pulse of current burns out the electronics in just the same manner as a surge of current in your home burns out a fuse instead of your oven or stereo. The difference is that the EMP attacks all electronics via a wave propagated in the atmosphere, unlike the surge of electricity in your home that goes through the wiring in the wall. You can easily protect against the home-brew pulse by simply installing a fuse box. But there is no fuse box possible when the entire circuit is under attack in all places at once. You have to shield the entire device that's to be protected, as noted in the statement above by Dr. Wood.

Certainly the most well-chronicled EMP in history came from an atmospheric nuclear blast over Johnston Island in the South Pacific in 1962 that was part of Project Starfish Prime. This 1.4-megaton explosion was set off at an altitude of about 250 miles over a remote area. But the resulting pulse of electromagnetic energy was felt in Honolulu, nearly seven hundred miles from the epicenter of the blast. Even though the energy in the pulse was considerably attenuated by the time it reached Hawaii, it still had enough "juice" to blow out streetlights in Honolulu, set off burglar alarms, and damage a communication relay station.

One should bear in mind that Project Starfish Prime was in 1962, nearly fifty years ago, when much of the world's electronics was still reliant upon vacuum-tube technology. In today's world of supersensitive microelectronics, all computers and cell phones; all cars, boats, planes, and trains; all critical infrastructures for supplying energy, food, water, and communication; and all electronic control and security systems are vulnerable. With this constellation of nice properties, an "EMP bomb" is truly a terrorist's best friend. But how easy is it to actually create such a pulse and deliver it over a wide geographic area?

To address this key question, we first must understand how the pulse is generated. An EMP begins with a short, intense burst of gamma rays of the sort produced by a nuclear explosion. It should be emphasized here that you do not need a nuclear blast to generate an EMP. But as the strength of the EMP goes up dramatically with the strength of the blast itself, you get a bigger "pulse for your buck" with a nuclear blast than with any other known type of explosive. I'll come back to this point later.

The gamma rays from the blast interact with air molecules in the atmosphere and scatter electrons at high energy in a process called the Compton effect. These high-energy electrons ionize the atmosphere, thus generating a very strong electric field. How strong depends on the magnitude of the blast, as well as on its altitude. The strongest EMP comes when the altitude is above twenty miles, but it's also very strong even with surface or low-altitude bursts. The weakest effect is when the burst is at an altitude somewhere in between.

To return briefly to a point raised earlier, it is not necessary to set off a nuclear blast to create an EMP. It can be done via conventional explosives and ordinary nineteenth-century physics by a device called a *flux compression generator (FCG)* or by a *magneto-hydrodynamic device (MHD)*. The FCG is just jargon terminology for a device that uses a quick-acting explosive to compress a magnetic field, transferring most of the energy in the explosive to the field.

An FCG consists of a tube packed with quick-reacting explosives. The tube is placed inside a slightly larger copper coil. Just prior to detonation, the coil is energized by a bank of capacitors in order to create a magnetic field. The detonation is then set off from the rear of the tube. As the electromagnetic wave expands outward from the force of the blast, the tube touches the coil creating a short circuit. The short moves forward as the tube flares outward, thereby compressing the magnetic field. According to Australian defense expert Carlo Kopp, "The result is that the FCG produces a ramping current pulse, which breaks before the device disintegrates from the explosion." This pulse has the strength of a million lightning strikes and is what burns out all electronics in the path of the electrical shock wave propagated outward from the FCG.

An MHD device operates on the slightly different principle of a conductor moving through a magnetic field, which then produces an electrical current perpendicular to the direction of the field and conductor motion. Frighteningly, either the FCG or MHD devices can be rigged up rather easily to serve as a very effective, compact, and cheap EMP bomb.

Regardless of how you create the EMP, the effects line up with the fictional works I discussed earlier. An instant after the bomb—nuclear, FCG, or MHD—is detonated, an invisible radio frequency wave is created. This "pulse" is more than a million times as powerful as the strongest radio signal from earthly radar, television, or radio sources. The wave is strong enough that it will reach every location in line of sight of the explosion. This is one of the principal reasons why a high-altitude burst can do so much damage. For instance, a single blast three hundred miles above Kansas will affect the entire United States, along with parts of Canada and Mexico!

When the electrical wave strikes the surface of the earth, it generates high-speed electromagnetic shock waves that threaten every part of our modern technological infrastructure, such as

- Computers and other devices containing microchip circuitry;
- All conductors and lines for electrical power transmission;

- All devices depending on electricity and electronics, ranging from security systems in banks to medical devices in hospitals to elevators in office buildings;
- And all cars, trains, planes, and boats.

So not only do all the electronics die, so does all the electrical power—perhaps permanently—since power transmission lines conduct the pulse to the transformers, which are then shorted out by voltages greater than what's seen in a typical lightning bolt.

What would you experience if you were unfortunate enough to be in range of an EMP attack? The first thing you'd notice is that lights, motors, elevators, and all other electrically powered devices stop dead in their tracks. Except for the immediate stoppage of moving vehicles like cars, trains, and planes, this would not be much different from the type of power failures many regions of the world have experienced numerous times before. Transportation systems wouldn't work, water wouldn't flow (since it requires electrical pumps to get the water to your faucet), and fluorescent lights and television sets would show a strange, otherworldly glow, even if they're turned off, due to the electron flux streaming through their noble gases or phosphors. Your smartphone would feel warm to the touch, its battery stretched well beyond its voltage limits. And, of course, your computer would be fried.

At first, you might think this was just another power failure—until you tried to access emergency communication channels to get the lowdown on what's happening. Those channels would be as dead as the rest of the communication systems we take for granted. Even if they're not, your battery-powered radio or cell phone will be. You'd be cut off from communication of all sorts other than direct verbal communication with those in your immediate vicinity. The vast majority of people would probably just break out the candles, fire up the backyard grill, and expect things to be back to "normal" in a few hours, or at most a day or two. But it ain't going to happen that way! If it's an EMP attack, the recovery time can be measured in many months, if not years. And by the end of the first week, panic would set in. Looters and hoarders would be out in force, law enforcement officials and the military would be deserting in droves to protect their own families, fires would burn unattended, and, in general, society would quickly revert to a lifestyle akin to the aftermath of a nuclear holocaust rather like that depicted in Cormac McCarthy's best-selling book and associated film *The Road*.

But unlike a nuclear attack, the EMP itself is totally harmless to humans. Unless you're dependent on technological helpers like dialysis machines, heart pacemakers, or other such electronic health-care devices, you'll survive the experience—at least for a while. This all sounds like the kind of weapon a mad scientist or equally deranged terrorist would salivate over. And maybe it is. Let's look a bit deeper at just how good a doomsday device an EMP bomb could really be.

NIGHTMARE WEAPON—OR JUST A BAD DREAM?

IS AN EMP BOMB A POOR MAN'S NUCLEAR WEAPON? IT'S CERTAINLY tempting to think that a weapon that's undetectable, untraceable, kills no people directly, totally devastates a modern society in a few milliseconds, and can be constructed with what amounts to 1940s technology is a great equalizer for a rogue state or a band of terrorists to flatten a nuclear power like the United States. And indeed, unverified reports suggest these features of the "E-bomb" have not been lost on the global terrorist community.

To illustrate, here is one possible terrorist scenario, based on actual facts, that would serve well for mounting an E-bomb attack:

- Al-Qaeda outfits one of its known fleet of freighters with a short-range ballistic-missile-launching capability.
- An anonymous SCUD missile or two is purchased from ready suppliers like North Korea. Cost? Less than one hundred thousand dollars.
- A modest-sized nuclear weapon is obtained from either a rogue

state like North Korea or Iran or from the large black-market inventory of weapons "missing" from the nuclear inventory of the former Soviet Union. Alternatively, a "no nuke," conventional explosives path along the lines sketched earlier is taken to create the E-bomb.

Given that countries like Iran have already demonstrated the ability to launch a SCUD missile from a ship at sea, these steps are all that's needed to get into the E-bomb business. The sea-based scenario sketched here would be especially attractive for a terrorist group since it may be very difficult to identify the attacker, given the vast quantity of SCUD missiles floating around the world today.

If you think this scenario is just the pipe dream of a slightly paranoid writer, think again. This is the very scenario presented in the report issued in 2004 by the congressional commission I mentioned at the beginning of this chapter. The commission reported that terrorists could totally disable the United States in one fell swoop by following precisely the steps of this scenario. A lone nuclear weapon launched from an offshore freighter is all it would take.

Of course, it's not at all a trivial matter to get your hands on a nuclear weapon, despite the fact that lots of them are unaccounted for and presumably in unfriendly hands. But as we've already seen, you don't need a nuclear weapon to create an EMP. A much simpler FCG or MHD device will serve nearly as well. But how likely is such an EMP scenario, nuclear or otherwise?

As reported in the *New York Times* in 1983, at that time strategic planners in both the United States and the Soviet Union regarded an EMP attack as the opening salvo in an all-out nuclear war. But it never happened, because the heart of nuclear defense in that era was the so-called MAD principle of "Mutually Assured Destruction," which served as a very effective deterrent to an EMP attack. This Cold War doctrine has pretty much been eliminated from strategic thinking today, though, as the geopolitical landscape has reconfigured itself.

Nowadays, the rise of nonstate players like al-Qaeda, combined with the widespread availability of weapons of mass destruction and the ever-shifting power balance between major nations, has made the possibility of an EMP attack much more difficult to evaluate. What does seem clear though is that these changes in the worldwide power landscape make an E-bomb threat much more credible.

The world now has many more nuclear states, some controlled by very unstable political regimes with legions of shadowy alliances—but without the capability of launching a full-scale nuclear war. So an EMP attack may well look rather attractive to such a state, especially if it could be carried out by one or another of the nonstate clients of this type of regime. As an attack of this sort will not destroy any lives and will not be followed up by an all-out nuclear strike, the country under attack is unlikely to launch a full-scale retaliation. In fact, it may well be impossible to know who the attacker actually is.

It's very difficult to prepare an "appropriate" response to an EMP attack. How do you respond to an explosion that takes place hundreds of miles up in the atmosphere without being seen or heard, yet destroys a whole national infrastructure in a moment? There are simply no legal precedents to give guidance in the formulation of a measured response to such an attack.

On balance, it would seem that an E-bomb attack has a very attractive cost-benefit ratio for the attacker. With one or two warheads you can devastate an entire country like the United States with minimal chance of retaliation. An EMP attack is also appealing as the opening salvo in a conventional war, since a state with a small number of warheads like North Korea or Iran may first want to take out the larger state's technological advantage prior to engaging on the battlefield by conventional means.

Su Tzu-yun, one of China's leading military analysts, stated the matter succinctly in 2001: "As soon as its computer networks come under attack and are destroyed, the country will slip into a state of paralysis and the lives of its people will grind to a halt." Even more ominous are the words of Iranian defense analyst Nashriyeh-e Siasi Nezami in 1999:

> . . . *Today when you disable a country's military high command through disruption of communications you will, in effect, disrupt all the affairs of that country . . . If the world's industrial countries fail to devise effective ways to defend themselves against dangerous electronic assaults, then they will disintegrate within a few years . . . American soldiers would not be able to find food to eat nor would they be able to fire a single shot.*

It may seem strange to most people to read what's stated here, since the natural question arises, "If an EMP is such a huge threat, why haven't I heard about it?" In late 2000, the US Congress tried valiantly to alert a recalcitrant White House to the danger by forming an EMP Threat Commission to study and report on the true magnitude of the EMP as a threat to national security. This was a counter to a report in 1997 from the Commission on Critical Infrastructure, which told the Congress that they regarded a terrorist attack via EMP as an event so unlikely that it didn't merit serious concern at that time.

Well, times change. And just seven years later in 2004 the EMP Threat Commission issued their report in which they stated that such a devastating attack was neither unlikely nor difficult to achieve. A member of that commission was the forementioned Dr. Lowell Wood, who testified that an EMP attack could send the United States back to a preindustrial age way of life in terms of society's ability to supply the population with food and water, not to mention cell phones, washing machines, pro football, and TV.

Before leaving the E-bomb weapon scenarios, it's worth pointing out that some respected scientists think the effect has been overrated. The EMP phenomenon has never been fully tested since the Comprehensive Test Ban Treaty, which came into effect very shortly after the Starfish Prime experiment, prevents nuclear tests in the atmosphere or in outer space. Consequently, the effects we have described here could end up being minor. The pulse may quickly dissipate due to distance, or there may be other factors that we don't know about like shielding from the effect by a mountain range that reduces its effect to more of a nuisance than a continent-wide catastrophe. These are the same sorts of unknowns that surrounded atomic weapons at the time of the Manhattan Project. The theory was there, but the application was lacking. From the blasts that destroyed Hiroshima and Nagasaki, now we know. One can hope we don't find out about an EMP in the same way.

I now rest the case for an E-bomb as a credible weapon of warfare and turn to the matter of the proverbial ounce of prevention instead of the kilos of cure. Supposing the EMP effect is truly real, not illusory, how might we protect ourselves from its devastation?

BEHIND THE WALL

AN EMP PULSE IS BOTH AN ELECTRIC AND A MAGNETIC FIELD, BUT it's the electric field that does the damage. While the pulse may last only a hundredth of a second, the field strength is so great that all exposed electrical equipment is likely to be destroyed. And not just electrical equipment. The EMP effect also disrupts the ionosphere, which affects the propagation of radio waves in many communication bands for up to a full day. Luckily, though, the amateur radio bands ("ham radio") would not be affected and could carry emergency communication without disruption.

The forms of damage that need protecting against come in two flavors:

1. ***Direct Damage:*** This is destruction of electronic components directly exposed to the EMP pulse shock wave. In this case, the protection must be something that prevents the electric field from reaching the exposed components.
2. ***Indirect Damage:*** Surges of electricity in an electrical line can arise from the EMP through an overload of current as the pulse

moves through the line. This type of damage can actually melt the long lines used for transmitting electrical power and telephone communication.

Protection against Type I damage involves isolating the equipment from the pulse by placing it inside a metal-covered, grounded box (what's called a "Faraday cage"). A possible Achilles' heel to this form of protection is the requirement that the equipment be *totally* isolated from the pulse. But since most electrical equipment has cables that connect it to things like power sockets or a communication link like a modem, just enclosing the equipment itself is no good. You must also insert surge protectors, spark gaps, or other forms of filters at the point of entry of the cable into the equipment to stop the power surge from reaching inside the protective box.

Preventing Type II damage in power lines calls for isolation of equipment and very detailed grounding, so that the electrical pulse has an easier time getting to the earth through the ground than through the equipment.

Unfortunately, hardening systems is difficult and expensive. Not only must equipment be encased in Faraday cages, windows must be coated with wire mesh and doors have to be sealed with conductive gaskets. Fortunately, fiber-optic cables are not susceptible to the EMP effect, so the move to replace copper cables with fiber optics will certainly contribute to reducing overall EMP vulnerability.

Of course, there are also indirect ways of protecting against an EMP by such means as maintaining backup units in shielded cages and keeping equipment out of the range of the pulse to begin with.

ADDING IT ALL UP

SINCE ALL EVIDENCE POINTS TO AN EMP AS BEING A CREDIBLE PHYSI-cal phenomenon, let's assume for the sake of exposition that the threat of an EMP attack is indeed real as described in the foregoing pages.

What about the timescale and likelihood of such an event actually coming to pass?

First of all, we can dispense with the notion of a naturally occurring EMP. To the best of current knowledge, the only way to create such a pulse is by human "ingenuity." It needs to be designed. Unlike many of the other X-events discussed in this book, nature is not in the business of throwing us an EMP or two just to keep things lively.

Given the ease of creating at least a low-level EMP device and the well-understood facts surrounding the havoc that such a device can wreak, we should consider ourselves lucky that no known EMP attacks have taken place to date. After all, there are a lot of disaffected groups scattered around the world, many with access to the type of technical skills and equipment needed to build at least an FCG or MHD device, if not a full-scale nuclear EMP generator. Perhaps the reason is similar to the arguments against using biological weapons: the impact is totally nondiscriminatory. The effect of the weapon destroys or contaminates the very region the attacker might want to control and make use of. A full-scale E-bomb can knock out the infrastructure of an entire society. No question about that. But destroying its infrastructure then makes the resources of that society largely unavailable to the attacker, as well.

Of course, not all attackers are created equal. And there's plenty of evidence of irrationality in the terrorist business to make the case that many potential attackers have no interest in taking over a society. Rather, they simply want to destroy it. For these sorts of attackers, an E-bomb would seem about as good a tool as one can find. Certainly, it would be a lot better than just blowing up a few buildings or nightclubs. Easy to build, dramatic in its effect, ensured anonymity of the attacker, relatively cheap to deliver—it's not difficult to imagine an EMP attack in today's highly charged, turbulent geopolitical climate.

A NEW WORLD DISORDER

THE COLLAPSE OF GLOBALIZATION

GOING GLOBAL

ONE OF THE BEST-SELLING BOOKS OF 2005 WAS *THE WORLD IS FLAT,* political columnist Thomas Friedman's account of the disappearance of national boundaries to the free flow of almost everything—money, people, labor, goods, ideas, and all the rest. In this award-winning volume, Friedman paints an evangelistic account of the phenomenon of globalization. Ironically, at just about the time the book was published, its central idea was already beginning to look much more like a publication from the Flat Earth Society than a visionary picture of how we will live in 2020. To Friedman's (partial) defense, though, in 2005 even professional futurists bought in to the notion of a "flat earth." As a small illustration of the groupthink prevalent at the time, I received a rousing silence accompanied by a sad, bemused shaking of heads from a group in Switzerland in 2006 when I had the temerity to deliver a presentation titled "The Decline and Fall of Globalization." An especially odd reaction, I thought, at a meeting of *futurists*! But so much for post postmortems. Let's fast-forward a few years and see how the future of globalization looks today.

Headlines in the financial press continue to paint an ever clearer picture of the emerging fact that the global financial system as it's currently constituted is incapable of dealing with the capital flows across borders that Friedman-style globalization demands. Initially, it seemed that globalization was mostly focused on the flow of jobs from regions like the United States where labor is expensive to China, Vietnam, and other places where it is cheap. But movement of jobs necessarily entails movement of the capital those jobs generate from the net importers back to the net exporters. These are the two pillars of the international trade and financial system that the flow of capital must balance. Unfortunately, that system is hopelessly broken.

Digging into the way in which people, money, and most everything else makes its way around the world, we see the specter of complexity hanging like a shroud over every step of the process. The system of globalization has given corporations a vast spectrum of ways (degrees of freedom) by which they can develop new products, produce existing products, market their wares, and the like through picking and choosing where and when these functions are carried out. So in a world with no national boundaries or constraints, transnational corporations have a huge level of complexity. On the other hand, the system composed of the global population at large, as represented by their national governments, has given up most of whatever freedom they had to regulate what can and cannot cross their borders without having to pay. In short, nations have voluntarily reduced their complexity in the business realm to a minimal level. As always, as that gap widened, so did the social stress of growing unemployment in Western countries, as all but the high-skill jobs moved to Asia. We are seeing the end result of this complexity mismatch in these very days, as the United States tries desperately to address the problems of a jobless recovery from the financial crash of 2007, while Europe struggles to deal with an even more serious financial crisis, not to mention the social disruption emerging from its own dangerously high levels of unemployment, particularly in the southern countries of the European Union like Greece, Italy, Spain, and Portugal.

As the ongoing tension between the United States and China dramatically illustrates, net exporters like China must accept an appreciation in the value of their currencies. On the other hand, net importers like the United States have to devalue. Of course, the exporting countries strenuously resist taking this step, as the revaluation process would bring the flow of goods and money into balance—precisely what the exporters do not want. Initially, this obvious fact gets played out in diplomatic circles. But if the diplomats don't get the job done within some acceptable period of time, financial markets will step in to do it for them. The result of that leveling of the playing field will not be pretty. In fact, this is another good example of a complexity mismatch that is very likely to be resolved by an X-event; namely, a massive devaluation of the US dollar, protectionist legislation, and a whole lot of other actions that will only accelerate the process of the world economy falling into a deep deflationary depression.

In an interesting juxtaposition of worldviews and timing, the year 2005 also saw publication of *The Collapse of Globalism*, a provocative work by Canadian polymath John Ralston Saul. While this volume garnered far less attention than Friedman's paean to the globalists, it's a much better guide to what we're seeing today and will continue to see in the decades to come. At its heart, Saul's work asks the question posed by a reviewer of the book, Michael Maiello: "Are political decisions meant to be made in deference to the economy and markets, or can we use our political institutions to shield us from some of the harsher effects that markets can dish out?" Adherents like Friedman claim that the powers of government will fade in the face of the power of markets. Saul thinks otherwise. According to the mythology of globalists like former Federal Reserve chairman Alan Greenspan, markets are self-regulating. But events like the Crash of 2007 show they most definitely are not. After more than three decades, globalization has failed in its promise to spread the wealth and reduce poverty. As noted by Pranab Bardhan in a 2006 article in *Scientific American* magazine, "Because the modern era of globalization has coincided with a sustained reduction in the proportion of people living in extreme poverty, one may conclude that globalization, on the whole, is not making the poor poorer. Equally, however, it cannot take much credit for the decrease in poverty, which in many cases preceded trade liberalization." Why should we imagine that these purported benefits will ever come from tinkering with local needs and concerns? In the end, the most important message from Saul's finely tuned argument is that the global economy is something that humans have created and is simply a part of human society. So shouldn't it serve our interests rather than trying to force us to serve its needs?

The retreat to localization has many faces that show themselves in different ways as we look at specific regions of the world. So let's have a few concrete examples to fix the underlying principle that an overdose of complexity can be bad for your economic and spiritual health.

A CATERPILLAR OR A BUTTERFLY?

MIKHAIL GORBACHEV'S RESIGNATION AS PRESIDENT OF THE USSR ON Christmas Day 1991 was a moment of hope for Russian liberals, who saw the dissolution of the Soviet Union as an opportunity for the Russian populace to join into the political, social, and economic life of the industrialized Western world. A population that had lived, loved, and labored at an abysmally low complexity level, with little freedom to travel abroad, choose leaders, or even buy consumer goods beyond the bare necessities of life was now empowered to do all three and more. Alas, fate decreed that Gorbachev hand over power to Prime Minister Boris Yeltsin, a hopeless drunk who almost immediately threw the reconstituted Russian Federation into political and economic chaos for the better part of eight years. By the time Yeltsin himself draped the mantle of power over the shoulders of former KGB agent Vladimir Putin at the end of 1999, the nascent movement toward a freer and more complex society had devolved into a land grab, in which

Yeltsin's cronies became today's "oligarchs" by effectively stripping all state assets worth taking and placing them in a few private hands (their own).

From the moment he took power in the 2000 election, Putin clamped down on any notion of political reform, free and open elections, public debate, and the like. Whatever complexity increase Russians acquired in the political arena was soon set back to its USSR level, where it remains to this day. Of course, an increase in the level of complexity in other areas of life like foreign travel, a semifree press, and consumerism was the trade-off Putin made for reducing the public's degrees of freedom in the political sphere. Some newly minted oligarchs, notably Russia's richest man at the time, Mikhail Khodorkovsky, didn't quite understand this conjuring trick, which called for a hard lesson from Mr. Putin to set him straight—in fact, straight to a prison camp in 2004 on trumped-up charges of tax evasion (shades of Al Capone's imprisonment in the USA on similar charges under similar circumstances in 1931).

A small but influential band of Russian liberals kept the faith during the interregnum of Dmitri Medvedev, who took power in 2008 when Putin was ineligible to run for a third term of office. In September 2011, one of those die-hard liberals, Lyubov Volkova, awoke the morning after Putin had "reappointed" himself president-to-be, saying in an interview with the *New York Times* that this development recalled a story similar to the one I told earlier in Part I about the butterfly flapping its wings in one part of the world, thus setting off a cascade of events that changes the world dramatically somewhere else. Ms. Volkova said, "Sometime—maybe not 20 years, but maybe 17 years ago—the butterfly was crushed and the consciousness of the Russian citizens traveled along a different path." Putin's return to the presidency will certainly stamp out whatever complexity gains in the political domain might have come from Medvedev's time in office, ensuring that the political complexity level of the Russian population will remain low for many years to come. So it seems open and free political expression in Russia wasn't a butterfly after all, eager to flap its wings and fly, but just a lowly caterpillar. (The mass protests in Russia in late 2011 seem to hold some promise for this butterfly, after all.)

This story of Russia's sad political plight illustrates two important complexity principles. The first is that the complexity level in a society can vary across different domains of life. Here we see the political complexity temporarily rising, but then rapidly quashed when the rise became too unsettling to the overall social order during the Yeltsin period. The attempts to modernize the country rested on an almost magical belief in the power of the free market. Liberalization led to the privatization of many state industries, which in turn created social unrest, bankruptcy of companies, an extremely high rate of unemployment, kidnappings, prostitution, and the rise of criminal groups similar to the Mafia in the United States in the Roaring Twenties. But at the same time the complexity level in other spheres of life, especially foreign travel and consumer goods, dramatically increased. This example is worth keeping in mind, as it suggests that governments can engineer trade-offs between different types of complexity to maintain their grip on political power. The situation in China today is another good example of just this sort of trade-off.

The second big idea, of course, is the butterfly effect referred to by Ms. Volkova. There was a moment when it actually looked as if a political butterfly was starting to emerge from its cocoon and flap its wings to set Russia off onto an entirely new course of political and economic freedom. But it was not to be.

THE DECLINE AND FALL OF THE EUROPEAN UNION

ANOTHER RECONFIGURATION, VERY DIFFERENT FROM THE SPECIFICS of the one in the former Soviet Union, yet eerily similar in other respects, is under way in Europe as I write. Here the entire edifice of the European Union is teetering on the edge of collapse, not through political and social unrest, but through what appears to be purely economic and financial causes. As money matters in every aspect of life, it's important for us to understand how this European crisis may unfold as a glimpse of how the world's geopolitical structure may look a few years from now.

Political analysts, editorial writers, and financial pundits, along with many other dreamers, schemers, and so-called men of affairs have pointed their collective finger at a wide variety of reasons for the financial quagmire the European Union finds itself bogged down in. These putative culprits run the spectrum from lazy Greeks, greedy bankers, and rapacious property developers to mindless Brussels technocrats and feckless politicians of every stripe. But such "explanations" are like doctors addressing the symptoms of a disease, rather than its cause. For the EU crisis, the cause runs much deeper than the mere caprices of an assortment of conceivably even well-meaning, but essentially clueless, individuals. The true *causa causarum* rests upon the fact that an ever-increasing gap in complexity between interacting human systems is almost inevitably brought back to contact with reality through "shock therapy." As I've repeatedly emphasized, this therapy generally takes the form of an X-event. Here's how those guiding principles work out in the context of the current EU crisis.

The formation of the EU can certainly be seen as a "joining," "globalizing" type of event. And indeed that event, the 1957 Treaty of Rome, took place at a time of increasingly strong feelings on the part of European governments that the time had come to unite into a single political body. Despite some setbacks in getting the EU Constitution approved in 2005–2008, the history of the EU has been pretty much onward and upward—until now! Forces of "separation" and "localization" have now begun to dominate, showing up in events ranging from the unwillingness of prosperous states in the EU to prop up the finances of the weaker members to talk of the reimposition of border controls by some countries to stem the flow of unwanted economic refugees from the Balkans, Turkey, and elsewhere.

As we've said earlier, when organizations, especially states or empires, encounter problems, the time-honored way to solve them is to add another layer of complexity onto the organization. Basically, this solution is the well-known process of "bureaucratic creep." As problems accumulate, the bloat of bureaucracy increases to the point where the entire resources of the organization are consumed in simply maintaining its current structure. When the next problem comes online, the organization falls off the cliff of complexity and simply collapses.

Many times this complexity trap shows up when two (or more) systems are in interaction. The gap in complexity between them becomes too big to sustain, and the resultant X-event emerges to close it. We saw this process earlier in the collapse of the repressive regimes in Tunisia and Egypt, both of which were facilitated by a rapid upgrading of the complexity of the low-complexity system, the citizenry of each country, via social networking and modern communication channels. The governments could neither suppress this buildup, nor keep up with it themselves. The end result was, of course, the X-event of rapid, violent regime change.

To illustrate this principle in the context of the EU, think of the Eurozone countries as a system in interaction with the rest of the global economy. If the countries were not in the Eurozone, they would have many options at their disposal to address changing economic times. They could, for instance, manage the supply of their own currencies, raise or lower interest rates, impose trade tariffs, and the like. In short, they would have a high level of complexity arising from the many different types of actions that could be taken.

Instead, the Eurozone members are severely constrained because no country can act unilaterally but must act in unison as per the dictates of the European Central Bank (ECB). So a complexity gap arises between a high-complexity system (the world) and a low-complexity one (the Eurozone states). Loans from the wealthier Eurozone countries to indebted ones and other efforts by the ECB to bridge this complexity gap will almost surely end up falling into the category of "throwing good money after bad." Ultimately, this will also almost surely lead to human nature's default solution for such a problem, which in this case will be the extreme event of a collapse of the euro, and very possibly the EU itself. Here it's no pun to call it a "default" solution.

Can the complexity gap be bridged without a collapse of the euro? Maybe it could have been, but only if the EU had taken a politically unpopular, but necessary, short-term "hit" when the debt crisis first arose instead of trying to buy their way out of a problem that money cannot solve. For example, implementing much more stringent regulatory procedures for vetting the finances of candidate member countries, or even slowing down the entire process for admission of new members, would have been painful and unpopular in the short run. But these sorts of actions would certainly have helped prevent the current crisis. This "faster is better" policy for admitting new member states was put in place to try to enlarge the EU as quickly as possible, presumably so that it would be "too big to fail." Talk about irony!

Even the policy of rapid expansion might have worked if it had been accompanied by the recognition that one-size-fits-all financial policies, while looking good in principle, almost always fail miserably in practice. Different cultures call for different approaches to almost everything, and to imagine that financial policies that work for a country like Germany could/should apply equally well to a country such as Greece or Portugal is only to invite a disaster.

Of course, it's now much too late in the day for actions of this sort. The historical record is filled with examples of ideologies that crashed into the brick wall of reality. The big question at the moment is whether the EU itself will end up in that graveyard of experiments in social engineering, an experiment that had to be done, but which is now likely to be seen as a noble failure. Let's be a bit more specific now about how things might look if the Eurozone does indeed come apart.

In times like today, many pundits argue that the future of the Eurozone rests primarily with its strongest member states, Germany and, to a lesser degree, France. The most likely way the Eurozone could fall would be if Germany saw the collapse of the Eurozone as being in its best interests. In short, the fundamental question is whether Germany would gain more by remaining in the Eurozone and, in essence, bankrolling it. Or would Germany's interest be best served by leaving? There are at least three major scenarios that might play out if

Germany decided for the latter course of action and left the Eurozone, each one of which constitutes its own special type of X-event.

Total Collapse: In this future, the Eurozone would revert back to what it was before the euro was introduced. This would require the European Central Bank to return its gold to member states of the Eurozone in proportion to their initial contribution. The myriad former national currencies—deutsche mark, lire, franc, Finnmark, and others—would be brought back to life and cross-values set to their levels at the time the euro was introduced.

In this scenario, the US dollar reserves built by the states will replace their euro reserves. Confidence in almost all currencies would disappear, as people and countries desperately seek hard assets like gold. It's safe to say that currency markets would be chaotic, probably with a strong initial move into the US dollar—but only until the smoke cleared away and business returned to something approximating normalcy.

Partial Collapse: It's more likely that the Eurozone will not totally collapse, at least initially, but will simply shrink by kicking out weak member countries from southern Europe. These southern countries—Spain, Greece, Portugal, Italy—would have to go back to their original currencies, which would take place in conjunction with imposition of currency controls to prevent flight capital from moving to the still-extant euro. The outcast countries would suffer years of poverty, but maybe not worse than what they will suffer by remaining in the Eurozone. The "new euro" would suddenly become the currency du jour, as the indebtedness of the remaining Eurozone countries would decline dramatically.

Unilateral Withdrawal: This is an extreme scenario in which the strongest member of the Eurozone decides it's had enough and sees that its national interest rests in flying solo. If this were to happen, the euro would be devastated to the advantage of the US dollar, which would remain the global reserve currency—but still fall slowly against other major currencies like the Japanese yen and Chinese renminbi.

Seers of the future love to spin scenarios like those above, laying a path to a new world order. I have to admit to a weakness for this sort of armchair philosophy myself—not so much to predict what will happen, but more to sketch a spectrum of possibilities for what *might* happen, ranging from the plausible (no X-events) to the highly speculative (many "high-X" X-events). Basically, scenarios are a way of focusing the mind both on constraints limiting how the world might look and on opportunities for shaping that world into something we'd prefer rather than a world we must simply endure. In this direction, let me tell a story about an exercise I participated in over a decade ago, but which still holds some very intriguing lessons for us as we try to understand the world as it might be a decade or two from now.

I WAS INVITED TO WASHINGTON, D.C., AT THE TURN OF THE MILLEN-nium to participate in a US government–sponsored exercise named Project Proteus. The goal of the project was to explore several very disparate scenarios for the way the world might look in the year 2020, and to evaluate what American interests would be threatened in such worlds and how to address those threats. The Proteus group consisted of about sixty people from a dazzling array of disciplines, ranging from physics and engineering to economics to science-fiction writing and even poetry. What's interesting for us is not the Proteus exercise itself, but some of the scenarios presented. (Tom Thomas of Deloitte Consulting and Michael Loescher of the Copernicus Institute created these visions of the world of 2020, and I'm indebted to them for their counsel and permission to reprint some of them here.)

Of the five worlds of 2020 presented to the group, the three worlds that seem most appropriate for our purposes here are those termed

Militant Shangri-La, The Enemy Within, and *Yankee Going Home.*
The full scenario for each world was incredibly detailed, consisting of
many, many pages of both data and fictional accounts of life in that
world. Following Thomas and Loescher, here is a telegraphic summary of each of these three worlds.

> ***Militant Shangri-La:*** This is a world of unexpected events
> and difficult-to-trace villains. The world, in general, and the
> United States, in particular, has continued into a third decade
> of a prosperous information-driven economy. But the world is
> also continuing along the road to complexity, with new struc-
> tures of influence on the globe. The Newtonian diplomatic and
> military calculus of the past four hundred years since nation-
> states emerged to close the Middle Ages seems to be giving way
> to another age. In particular, the global man-in-the-street has
> endured the past century of two hundred million deaths in war,
> endured dizzying and difficult technological change, and is lis-
> tening sympathetically to the earth groan under the burden of
> population and extinction. Nearly all of the animals of Africa,
> many of fish in the sea, and much of the wild areas of the globe
> are used up. Into this world enters the new and worrisome
> Alliance of the Southern Constellation: South Africa, India,
> Indonesia, China, and other pariahs to the Western social phi-
> losophy of individual liberty and human rights, operating both
> legitimately as a block of aligned nation-states and illegitimately
> as criminal cartels. Their grand strategy is to keep the world on
> the edge of chaos and, from that chaos, reap profit. The Alli-
> ance is in space, on the seas, in the media and financial institu-
> tions, and worming into the hearts and minds of individuals
> to kill the very idea of personal liberty. Meanwhile, the United
> States, its four English-speaking cousins, and their Pacific allies
> Japan and a newly unified Korea unite to resist the evil empire.
> ***The Enemy Within:*** This is a world in which the United
> States has slowly and unexpectedly, but quite dramati-

cally, unraveled. Like so many other nations at the height of power, disagreements, ethnic tensions, and single-issue politics have torn the social fabric. The society is fractured and fragmented—politically, socially, and culturally. Intergenerational strife, compounded by record unemployment, has torn apart churches, neighborhoods, and families. Racial tensions are a tinderbox in both cities and suburbs. The United States has become an uncivil society, with uncertainty the specter looming over each activity of every day. Violence can pop up at any time and in the most unlikely places. There seems to be no refuge. Under such social circumstances, capital and business are flowing out of the country. The nation's economy creaks along at barely sustainable levels. Agriculture, health care and pharmaceuticals, low-end retail, personal security services, and construction are among the few bright spots in this abysmal economy. Government coalitions struggle to find an appropriate national response to the seemingly never-ending crisis. All other national tasks and obligations are deemed unimportant as the country turns inward and faces the most critical turning point of its 250-year history.

Yankee Going Home: This is a world in which little is clear except that the world has changed in fundamental ways. Who is running things? Why are certain decisions being made? What goals are being pursued? Who are friends and who are enemies? The United States has withdrawn from the world, gone home after a series of terrible foreign-policy blunders and after a long-standing and deep recession. The world is heavily influenced by the memories of terrorism, regional war, and worldwide instability that followed the US isolationism. In the wake of the US retreat, we find a world made up of both traditional actors (nations, international organizations, nongovernmental organizations) and powerful nontraditional actors (global corporate alliances, criminal groups, mercenary units).

These actors cooperate for power and influence while simultaneously competing for position and control in a constant whirl of politics and economics, bewildering to nearly all concerned. In this world, historical notions of allegiances are questioned, and the rules of the game are difficult to understand. Predictable behavior becomes the unique exception rather than the expected standard.

The Proteus scenarios show how the United States might decline, if not fade away altogether, as a world power. It's interesting to see how these paths to ignominy match up with visions of the end of the American empire set out by another famed visionary.

PETER SCHWARTZ IS PROBABLY THE WORLD'S MOST WELL-KNOWN FUturist. Formerly head of scenario planning for Royal Dutch Shell, he founded the Global Business Network (GBN) some years ago in order to explore various scenarios for the future for clients ranging from the US Department of Defense to filmmaker Steven Spielberg for his production *Minority Report*. In August 2009, Schwartz was approached by *Slate* magazine to create alternative visions for how the United States might leave the world's geopolitical center stage sometime in the next hundred years. His group developed four different paths by which this might occur. Here is a brief summary of each of the roads to collapse that the GBN group cooked up.

Collapse: Following government-bungled responses to a series of Hurricane Katrina–like catastrophes, the mood of the American population flips to the negative, as the public begins to see their own government as the common enemy. This shift in the collective psychology of the population results in a complexity mismatch between the government and the public, much like what's happened recently in the Arab countries of north Africa, the end result being an implosion of the United States due to unsustainable internal divisions.

Friendly Breakup: This scenario involves a downsizing in which the United States breaks apart due simply to an inability to bear the cost of running a huge empire. Schwartz sees this dissolution as analogous to the breakup of the Soviet Union. A variation of this scenario would be if a big state, say California or Texas, or a region like the West Coast would develop sufficient resources and individuality to split off from the union. The GBN argues that one way this could happen would be for the politically left-leaning states to join together into a "Democratic Alliance," while states at the opposite end of the political spectrum would form a "Republican Nation."

Global Governance: In this world, the United States gradually declines in geopolitical significance as it is assimilated into a larger global community. In short, the world bands together to form a true "United Nations," and all nation-states, including the United States, cede much of their authority to that overarching global government.

Global Conquest: This is the hard path to the sidelines, in which not only the United States but also the rest of the world is subjugated to a world dictatorship. A kind of "super Mao," as Schwartz terms the dictator, takes over the world by force, probably using space-based weaponry, and shuts down the rest of the world.

The three Proteus pictures of the world of 2020, as well as the GBN scenarios, illustrate the way scenarios might be used for anticipating the future. As we see from recent events, each of them contains pieces of the actual world as it seems to be unfolding, and each gives a glimpse of how an X-event is likely to impact what we'd see a decade from now. Both the Proteus organizers and the Schwartz group caution that the scenarios should not be taken as predictions of the future, but more as thought experiments to stimulate discussion about the various factors that might combine to create such an event. In short, we have to mix and match from each of the scenarios in order to come up with something that would look to be the way to bet today on the most likely world of tomorrow.

With these examples of Russia, the EU, and Project Proteus under our belts, I'll wrap things up by speaking briefly about the strategic aspects of historical cycles and complexity gaps, together with how as individuals we might weather the economic, political, and social storms forming on the visible horizon.

All the Proteus scenarios envision social collapse proceeding as a slow train wreck, a gradual, almost leisurely, process whereby one social group (society, empire, civilization) smoothly passes on the baton of global power and influence to its successor. Of course, this "passing on" is more in the order of the newly minted power yanking the baton away from the ancien régime than a friendly handoff. Nevertheless, the cyclical theories of historical processes espoused by twentieth-century thinkers, ranging from Oswald Spengler and Arnold Toynbee to Paul Kennedy, all see such a smooth transition. Basically, the received wisdom is that history has a rhythm and that that rhythm entails gradual change with no huge discontinuities. Recently, Harvard historian and general social thinker Niall Ferguson has argued a very different picture for how this transition really takes place. A brief account of Ferguson's ideas is a good place to begin our summing up.

FITS AND SPURTS

SOME YEARS AGO, EVOLUTIONARY BIOLOGISTS STEPHEN J. GOULD and Nils Eldredge put forth a theory of evolutionary processes they termed "punctuated equilibrium." Their claim was that evolutionary processes do not take place in a slow, gradual fashion but rather occur in fits and spurts. For long periods of evolutionary time (hundreds of thousands or even many millions of years), more or less nothing happens. Then along comes a period like the Cambrian explosion, which took place about 650 million years ago, during which a huge number of dramatic evolutionary changes take place. In a short (by evolutionary standards) period of 510 million years, the major animal groups we know today appeared, animals with shells and external skeletons. Afterward, things settled back into a kind of long-term "hibernation."

Niall Ferguson's view of the dynamics of historical processes is very reminiscent of the Gould-Eldredge theory for biological processes. And why not? After all, history is itself a social process involving evolutionary change. So it's not unreasonable to imagine that the biological mechanisms of change and the historical ones (whatever they may turn out to be) would show great similarities.

What Ferguson has in mind for historical change is a process that will gladden the heart of any system theorist. Ferguson explained his radically different view of the way history unfolds in a 2010 article in the journal *Foreign Affairs*:

> *Great powers are, I would suggest, complex systems, made up of a very large number of interacting components that are asymmetrically organized. . . . They operate somewhere between order and disorder—on the "edge of chaos," . . . Such systems can appear to operate quite stably for some time; they seem to be in equilibrium but are, in fact, constantly adapting. But there comes a moment when complex systems "go critical." A very small trigger can set off a "phase transition" from a benign equilibrium to a crisis. . . .*

So there it is: the Gould-Eldredge punctuated equilibrium theory scaled up (or down!) to the domain of social processes.

In his argument, Ferguson states that any large-scale political unit is a complex system, be it a dictatorship or a democracy. In particular, empires show the characteristic tendency of a complex system to move from stability to instability very rapidly. Cyclical theories of history have no room for such acyclical discontinuities, which is perhaps not surprising as the theory of complex systems is a relatively new entry into the modeling pantheon, having emerged in full force only over the past few decades.

Ferguson buttresses his argument with numerous historical examples of empires that dropped off the edge almost overnight rather than doing a slow fade into the sunset of history. It's of considerable interest to note the benchmark case of the Roman Empire, which collapsed in just two generations, with the city of Rome itself suffering a population decrease of nearly 75 percent during that time. The archaeological evidence—lower-grade housing, fewer coins, smaller cattle—shows the downsizing phenomenon I've noted several times in this volume and reflects a dramatic reduction in the influence of Rome on the rest of Europe. A more recent example is the fall of the Soviet Union in 1989, which I discussed earlier in this chapter. The historical record traces many more sudden empire "crashes" in between these two.

What does this all suggest for the prospects of the United States in the years to come?

As Ferguson notes, empire transitions happen virtually overnight. As a result, it's largely a waste of time to talk about stages of decline and wonder where the United States stands today in that transition. Moreover, most empires ultimately fall due to financial mismanagement and their attendant crises. In essence, the gap between income and expenses widens precipitously, and the empire is ultimately unable to finance this debt (yet one more complexity gap). Looking at the escalation of US public debt from $5.8 trillion in 2008 to a projected $14.3 trillion a decade or so from now should be enough to throw fear of the fiscal gods into anyone.

As I noted in Part I, a driving factor in what actually happens in the social realm is the beliefs people hold about their future, the so-called social mood. As long as people believe the United States will be able to deal with its problems, all is well and the country, along with the rest of the world, will make it through whatever the crisis of the day may be. But the moment a financial butterfly flaps its wings in the form of a seemingly innocuous event, perhaps the failure of a bank (like Lehman Brothers in 2007) or the downgrading of a minor (or major) country's debt (like the USA in 2011 or France in 2012), the whole house of cards can come tumbling down as both panicked investors and the general public start running for the exits. As Ferguson states it, the system "is in big trouble when its component parts lose faith in its viability." The conclusion from this argument is that empires function for some unpredictable period in seeming equilibrium—and then they fail abruptly.

Now we ask how all these fancy abstractions and general principles translate into the kind of life Americans will probably be living when a deflation-fed depression or a hyperinflation sets in for real? Here's a short preview of coming distractions. How likely are any of the scenarios I've presented to actually occur? From today's perspective, none of them look especially likely, particularly if you're a trend follower. But a few decades is a long time, and there will certainly be many surprises between now and then. One need only consider things like nanoscale weaponry, catastrophic weather pattern changes, a new Ice Age, or almost any of the X-events I've put on the table in this section of the book. Any of these might well dramatically shift the odds in the coming decades. Moreover, if Niall Ferguson is right, the time to prepare is now, since the collapse, if/when it comes, will take place quickly and by then it will already be too late.

ADDING IT ALL UP

A BLOOMBERG HEADLINE CAUGHT MY EYE THE OTHER DAY AS IT PRO-claimed, "Apocalypse Angst Adds to Terrorist Threat as Rich Russians Acquire Bunkers." The story went on to tell about a firm building $400,000 private bunkers for oligarchs at hidden locations across Russia, in order to protect themselves against the global cataclysm foretold by the ancient Mayan calendar for the end of 2012. For those with a less-than-oligarch-level bank balance, another firm is building communal bunkers at undisclosed locations in central Europe, where you can check in for $25,000 per person when it all comes undone. While it seems unlikely that even a figurative bombshell X-event would call for adoption of literally a bunker mentality to survive the fallout, there can be little doubt that today's post-postindustrial lifestyle will be in for some serious downsizing (maybe even for oligarchs) if any of the X-events presented here were actually to take place.

As a somewhat smaller step in the direction of lifestyle change, a survivalist-oriented website, www.usacarry.com, contains a very entertaining, and even possibly useful, article titled "The Top Ten Survival Things You Need to Do Before the Complete Collapse of the U.S. Dollar." I was certainly eager to see how the author, a certain Mr. Jason Hanson, saw life without the US dollar, so I looked in on his site. He wrote, "There will be riots in the streets and Marshall [sic] Law," as part of the American scene. As for his ten steps toward ensuring a place in the new America, the first thing to do according to Mr. Hanson is to have "at the very minimum the following three guns: A handgun, a rifle and a shotgun." Later, after laying in a year's supply of food and a month's worth of water, the article finally comes around to money: some gold, silver, and cash. So there it is. The article concludes with the admonition: "And most importantly, get those guns!"

Well, OK. If you're ready to live in a world where the survivors will probably envy the dead, an armload of weapons and an underground bunker might be just the ticket. But an Armageddon-like post-nuclear-holocaust type of world versus what we're likely to face in a post-US dollar, deflationary world are very probably going to be quite different things. For a bit of perspective, think about how the current Great Recession has already affected family life and then just scale it up to a Greater Depression level.

According to a mid-2010 survey by the Pew Foundation, more than half of all adults in the US labor force had experienced some "work-related hardship," such as prolonged unemployment or reduced work hours since the beginning of the recession in late 2007. Another survey showed that more than 70 percent of Americans over the age of forty felt affected by the economic crisis, and that the net worth of the average American household declined by around 20 percent. So the impact of simply a major recession has already put lasting dents in the American way of life.

The other side of this bleak economic coin is the social benefits that many believe will accrue from being forced to adopt a less profligate way of life. Myth has it that the Great Depression of the 1930s was ultimately redemptive, by forcing society to work together to bring a sense of unity to the country. But reality is a harsh mistress, and the facts of the matter are that it was the Second World War that served this function, not the Great Depression. At present, the Great Recession is showing no signs of bringing about a simpler, slower, less consumer-oriented way of life. Most people are simply getting poorer as the rich get richer, while family relationships are strained and in many cases coming totally unglued. It's difficult to imagine that a total collapse of the global economy would make this picture any brighter.

Instead of a brighter, better America, a postcrash world is almost surely destined to reset what we mean by "normal." Here are just a handful of the "new normals" that *Fortune* magazine identifies as being what we're likely to see:

Renting, Not Owning: The central pillar upon which the so-called American Dream rests is home ownership. Having your own piece of land and a home as your castle goes right along with Mom and apple pie. While Mom and apple pie may survive as a part of the postcrash America, home ownership will definitely have to go. The rich will own; the rest will rent.

Permanent Unemployment: The US economy would have to add more than 300,000 new jobs per month over the next three years to get the jobless rate below 7 percent by 2014. Nowadays, a monthly report that adds fewer than 100,000 jobs is heralded as a major advance. So a return to the pre-

2007 levels of 5 percent unemployment or lower is a distant dream. And it won't get any closer when the US economy steps off center stage.

Saving, Not Buying: Reduced income and uncertainty surrounding jobs translates into paying down debt and saving for arrival of the pink slip, not picking up a nice, but unnecessary, pair of glitzy shoes or treating yourself to dinner at a fancy French restaurant.

Higher Taxes for "the Rich": According to today's lexicon, "rich" means earning $250,000 a year or more. Where that figure comes from is anyone's guess. But somehow it's been enshrined in Washington-speak to mean the boundary between those who should pay more income tax and those who shouldn't. This sounds great. After all, less than 2 percent of US households have an annual income this high. But, in fact, if the dollar collapses and hyperinflation sets in, that quarter of a mil isn't going to buy so much anymore. In fact, some folks say it doesn't even buy that much right now. On the other hand, with a scenario of deflation that does not flip into hyperinflation, that money will go a very long way indeed—provided you've buried your cash in the backyard or stuffed it into a mattress so you can get your hands on it when your bank goes belly-up.

In summary, the way of life in the industrialized world will definitely take on a bleaker and more somber tone when the world separates into a cluster of localized, not globalized, power blocs. But Armageddon it ain't (probably)!

DEATH BY PHYSICS

DESTRUCTION OF THE EARTH THROUGH THE CREATION OF EXOTIC PARTICLES

KILLING THE EXPERIMENTER

IN MY HIGH SCHOOL CHEMISTRY CLASS, THE MOST EXCITING MOMENTS always came when the teacher performed an "explosive" experiment to get the class's attention. One such experiment that I recall fondly involved slicing a bit of metallic sodium from a bar of sodium immersed in a murky fluid and then dropping it into a flask of water. The reaction immediately separated the water molecules into their components, hydrogen and oxygen. It also generated a lot of heat. What made the experiment memorable was that the teacher obviously miscalculated the amount of sodium to be used, as the heat in the reaction, together with the oxygen, ignited the hydrogen, leading to a big explosion that broke the glass containing the water and put a huge circular burn mark on the ceiling of the lecture hall. Luckily, no one was hurt in this experiment gone awry. But it was the end of that type of attention getter for the remainder of the semester and serves well to illustrate the idea of an experiment that might easily have killed the experimenter.

While small potatoes as an existential threat, a high school chemistry lab experiment like this going astray is a good example of how complexity can rise up to bite you with potentially disastrous consequences when you're not looking. In this case, we see the butterfly effect in action whereby a small miscalculation on the part of the teacher as to the amount of sodium to be used in the experiment led to a runaway reaction that didn't quite either blow up the lab or kill the teacher but quickly sobered up everyone in the room.

So just like the ice-nine experiment in Vonnegut's book *Cat's Cradle,* which I detailed in Part I, playing with forces of nature that you don't quite understand can be very dangerous to not only your own health, but to that of everyone on the planet should things go totally off the track.

Another such experiment on a vastly greater scale was the first atomic bomb test at Trinity site near Alamogordo, New Mexico, on July 16, 1945. As early as summer of 1942 in Los Alamos, Edward Teller, one of the scientists developing the bomb, expressed concern that the enormous temperatures generated by the explosion might set fire to the earth's atmosphere. Just envisioning this gigantic, two-hundred-yard-wide mushroom cloud might well convince a person to take seriously the idea that the entire planet might just possibly be consumed in such a monumental fireball.

Despite the fact that the blast would generate temperatures hotter than those at the core of the sun, most of Teller's colleagues felt his idea of a self-sustaining fire being ignited in the atmosphere was a very remote possibility. The director of the Manhattan Project, J. Robert Oppenheimer, called for a study of the matter. The report, publicly available only since 1973, confirmed the skeptics' view that a nuclear fireball cools down far too rapidly to set the atmosphere aflame. But there was another threat hidden in that test.

Little was known about the dangers of radiation exposure in the

1940s, so local residents near the Trinity site were not warned or evacuated in advance of—or even following—the test. As a result, people in surrounding areas were exposed to radiation by breathing contaminated air, eating contaminated foods, and drinking affected water and milk. Some ranches were located within fifteen miles of ground zero, and commercial crops were grown nearby. At some of these ranches, exposure rates were measured shortly after the blast showing levels of about 15,000 millirems per hour, more than ten thousand times greater than what's now considered to be safe. Even today, a one-hour visit to the Trinity site will lead to an exposure level of 0.5 to 1.0 millirems, which is about the amount of radiation a typical adult receives daily from natural and human sources, like X-rays and radioactive elements in the soil.

Physicists expressed concerns of a similar nature when the first sustained nuclear reaction was established in December 1942 by Enrico Fermi's group under the abandoned west stands of the football stadium at the University of Chicago. Fermi had convinced the scientists that there could be no runaway nuclear reaction and that the city of Chicago was "safe." Nevertheless, historians from the Atomic Energy Commission noted that it was still a "gamble" to conduct such an experiment with totally untested technology at the heart of one of the country's largest cities.

Even though the nuclear test in New Mexico did not really threaten life on Earth, at least not by burning up the atmosphere, the consideration that that might happen is the first time in history that scientists seriously looked at whether their work might destroy the planet. As technology has moved forward at an ever-accelerating pace, these same sorts of fears continually appear and reappear. Their most recent manifestation is that the planet might be sucked into a man-made black hole or disappear in a shower of even stranger particles emerging from the gigantic particle accelerators at Brookhaven National Laboratory in the United States and at the European Nuclear Research Center (CERN) on the Swiss-French border outside Geneva. The basic question that arises whenever a new machine is built, generating ever more violent collisions of the elementary particles circulating inside its rings, is whether those violent collisions could spawn some kind of particle or event that would somehow "vacuum up" the earth—or even the entire universe. In particular, the fears circulating around the Large Hadron Collider (LHC) that went into service at CERN in late 2009, were that a particular form of a really strange particle aptly called a "strangelet" would just appear and a moment later the earth would just disappear.

Before digging just a bit deeper into why some feared such an outcome from the LHC, it's of more than passing interest to examine why we build such potentially dangerous and definitely expensive "toys" like these particle accelerators in the first place. They are without doubt the most expensive laboratories ever created, and they represent the leading edge of technology. So what are we expecting to gain from such a gigantic investment in brainpower, engineering dexterity, and just plain hard cash?

SOMETHING—OR NOTHING?

THE 1960S WAS AN ESPECIALLY ACTIVE DECADE FOR THEORETICAL physicists advancing models to encompass all that was known about matter, energy, and everything else. Further development of this work has led to what today is termed the "theory of everything," which is meant to embrace in one compact mathematical theory the workings of the many particles and forces that govern the universe, explaining how it began and how it will end. But there is still a missing link in this so-called Standard Model, an as-yet-unobserved elementary particle called the *Higgs boson*, which explains how matter comes to have its mass.

When British physicist Peter Higgs postulated such a particle in the early 1960s, his suggestion was met with derision by most of his colleagues. Today, the betting odds are that one of the triumphant products of the LHC will be the first actual observation of this elusive object. If CERN physicists can really find the Higgs boson, the Standard Model that the vast majority of physicists believe in today will be confirmed—and Higgs himself, who is now pushing eighty years of age, will have his crowning moment of personal and professional vindication.

Higgs developed his theory to explain why mass disappears as matter is broken down to its elementary constituents. His theory claims that at the very moment of the Big Bang, matter had no mass at all. And then it instantly gained it. The question is, How did this process work? Higgs argued that the mass must be due to an energy field that clung to the particles as they passed through the field of the Higgs particle and that gave them mass. That mysterious particle is now often termed the "God particle," a label Higgs himself dismisses, especially as he professes to be an atheist. But without such a particle, stars and planets would never have formed since the matter created at the Big Bang would simply have moved off into space and never gravitated together to form massive objects, or for that matter, into organisms like you and me.

So confirmation of the existence of the Higgs boson is the first order of priority for the LHC. But scientists caution that even if the God particle is actually there, we may not see it. The process by which the particle gives mass to matter happens so fast that it may be buried in the data collected from the LHC and could take many years of "data mining" to find.

But the Higgs boson is not the only treasure that may pop up from the LHC once it's up and running at full strength. Another possibility is that the collider will turn up evidence supporting the most theoretical of theoretical ideas in modern physics, string theory. There is a vocal community in the world of physics who argue that the entire universe consists of ultramicroscopic "strings" of matter-energy. That's it. It's strings of one type or another that form the entire universe as we know it. The problem is that no one has ever found a single solid piece of experimental evidence to support this theory! The whole notion is pure mathematical speculation.

To make any of the string theories work requires the universe to possess unseen dimensions beyond the normal three dimensions of space and one dimension of time that we're familiar with from everyday experience. Most string theorists believe in a world of ten dimensions, and they hope that the LHC will uncover those extra dimensions. How might this happen?

One way the LHC might establish the existence of new dimensions would be if it creates micro black holes. The decay rates of the subatomic particles that such a black hole creates can be analyzed to see whether hidden dimensions actually exist. A closely related way to establish the existence of these "missing" dimensions would be for the LHC to produce gravitons, particles that carry the gravitational force, disappearing into these other dimensions. Anything of this sort would be nothing but sweet music to the ears of string theorists, lending some actual experimental evidence to their flights of mathematical fancy. Preliminary results, though, look distinctly unpromising.

At a physics meeting in Mumbai in late summer 2011, experiments were presented by Dr. Tara Spears of CERN, who stated that researchers failed to find evidence of so-called supersymmetric particles. This result puts one of the most popular theories in physics, superstring theory, on the spot. If the conclusions presented by Spears hold up, then physicists are going to have to find a new "theory of everything." Interestingly, earlier results from the Tevatron in Chicago suggested just the opposite, which is why researchers asked CERN to employ the LHC to examine the process in more detail. Professor Jordan Nash of Imperial College in London, one of the researchers on the CERN project, states the matter this way: "The fact that we haven't seen any evidence of it [supersymmetry] tells us that either our understanding of it is incomplete, or it's a little different to what we thought—or maybe it doesn't exist at all." Before declaring supersymmetry a dead duck, though, we have to bear in mind that there are many other versions of the theory, albeit more complicated ones, that have not been ruled out by the LHC results. Superparticles may just be a lot harder to find than physicists originally thought.

As an interesting glimpse into the sociology of science, the downfall of supersymmetry would be a heavenly vision to a generation of younger theoretical physicists, who would then find the field wide open again for them to invent new theories, rather than being saddled with something invented by the older generation. As Max Planck once stated the matter, new theories are never accepted, they just have to wait for their opponents to die off. In this case, the "opponent" may well turn out to be supersymmetry. The next few years should pretty much settle the matter. But possibly there's other bounty to be obtained from the LHC besides supersymmetry.

Probably the most puzzling fact we've observed about the universe as we see it is that there simply doesn't appear to be nearly enough visible objects—stars, planets, asteroids, and so on—to account for the gravitational forces holding the galaxies and the universe itself together; to do the job, there must be a lot more gravity-generating matter than we currently observe. Enter "dark matter," a form of matter that can't be seen but makes up much more of the universe than all the visible matter taken together.

If (and it's a huge *if*) dark matter exists and has the right interaction strength with visible matter, some theories predict that the particles produced by collisions in the LHC will decay into dark matter that could actually be observed. But it's not known whether it's even possible to create dark matter by putting enough energy into a small enough space, so it might not show up in a collider yet more powerful than the LHC. And if it does appear, we know so little about the properties of such matter that it could be there and we'll never see it because we don't really know how to look. The only thing physicists seem confident about is that if it exists, it interacts very weakly with known particles. This means it would be hard to separate dark matter from background noise in the LHC experiments. So it's a long shot, at best. But if the LHC can create particles that are even good candidates for being dark matter, that would lend encouragement to the whole idea. Last, but not least, there are the strangelets.

• • •

In 1993, two mysterious explosions shot through Earth at nearly a million miles an hour. Whatever the objects were, on October 22 they set off earthquake detectors in Turkey and Bolivia that noted an explosion in Antarctica packing a wallop of several thousand tons of TNT. Just twenty-six seconds later, whatever that object was exited the floor of the Indian Ocean near Sri Lanka. One month later on November 24, a second event was detected. Sensors in Australia and Bolivia showed an explosion off the coast of the Pitcairn Islands in the South Pacific, and an exit of the object in Antarctica nineteen seconds later.

According to physicists, both explosions are consistent with an impact by strangelets, bizarre particles that are postulated to have been created during the Big Bang and are still being formed inside very dense stars. Unlike ordinary matter, though, strangelets contain "strange" quarks, particles that are normally seen only in the shower of particles generated inside huge particle accelerators. The team investigating the 1993 events says that two strangelets just one-tenth the width of a human hair could account for the observations.

Smashing protons together at the energies reached by the LHC could create new combinations of quarks, the particles that form protons. It's just possible that the kind of strange quark(s) that make up a strangelet will be produced in these collisions, as well.

The impact of a mini black hole created in a particle accelerator like the LHC has been explored extensively in the so-called hard science-fiction literature, since that's the only place physicists can give expression to their imagined fears of an event that to the best of our knowledge has not yet occurred. Unfortunately, these Hollywood-style venues for exposition of the effects of a mini black hole are strongly at odds with the reality of what we actually know about such objects, assuming they even exist.

To cut to the chase, here is the situation with mini black holes as we understand them today:

1. In order for such an object to be produced, the extra dimensions discussed above would have to exist; moreover,

2. if the mini black hole did not immediately evaporate as expected (through what's termed "Hawking evaporation," a very strong prediction by the famed physicist Stephen Hawking), then

3. the velocity of most of the mini black holes would be so great that they will escape the earth's gravitational field permanently. In the very rare instance that one of these ultra-rapidly-moving objects struck a proton or neutron inside the earth, the momentum of the black hole would still not be altered in any important way.

So the bottom line is that mini black holes are not any kind of threat to humankind.

Turning to the question of dark matter, the impact in the short term would be simply one of aesthetics. Our theories of the universe as we know it today require a lot more matter than what's been seen. Discovery of this "missing" matter would tidy up our theories and give us some confidence in forecasting the ultimate fate of the universe. Either the current expansion will continue indefinitely, or there will eventually be a Big Crunch consisting of a contraction back to a single point. Finally, the third, but unlikely possibility, is a steady state in which things are balanced just right so that there is no cosmic oscillation between the Big Bang and the Big Crunch, but only a Big Yawn. So there is no immediate threat to our human way of life from dark matter, either.

That's the menu: mini black holes, strangelets, dark matter, the Higgs boson, hidden dimensions. And those are just the things that physicists know about or postulate that might turn up in the debris of the LHC collisions. Detection of one or another of these objects would validate one model of particle physics over the competitors.

The real prize, though, would be something that we don't know about—a kind of Unknown Unknown! If such an event should occur, the entire world might vanish, and with that disappearance the world of physics, along with everything else. Or maybe such an X-event would just turn the world of physics upside down and force us to rethink everything we believe we know about the mysterious ways of the material world.

The opposite end of this spectrum is the booby prize: We find nothing! Years of smashing particles together yield nothing that we don't already know about. Should this occur, we would probably also have to rethink our theories of the universe. So the two extremes, something totally new or just plain nothing, might end up being the most exciting discovery of all.

FEAR OF PHYSICS

IN ITS MARCH 1999 ISSUE, THE POPULAR-SCIENCE MAGAZINE *SCIENTIFIC American* ran an article titled "A Little Big Bang," which signaled the second phase of concern that physics would/could destroy the planet, if not the entire universe. The focus of concern in that article was that the Relativistic Heavy Ion Collider (RHIC) at the Brookhaven National Laboratory on Long Island outside New York City would create strange particles of matter that might either blow up the planet or perhaps suck the entire universe into a hole from which it would never return.

The RHIC consists of two circular tubes, 2.4 miles long. Electrons from gold atoms are isolated and then accelerated to 99.9 percent of the speed of light. When these electrons collide, incredibly dense matter is created having a temperature ten thousand times greater than at the center of the sun. These are conditions that haven't existed since the creation of the universe in the Big Bang, a moment twelve billion years ago when all the laws of physics known today broke down. So it's natural to wonder what the effects might be of such an experiment. In fact, that very question is the reason the RHIC was built in the first place.

Following the *Scientific American* article, numerous letters from concerned readers flooded into the magazine's offices expressing anxieties over the possibilities that the collider would destroy us all. Typical was the letter from Mr. Walter Wagner, a former nuclear safety engineer turned botanist in Hawaii, who said Stephen Hawking's work argued that a miniature black hole would have been created moments following the Big Bang that started the universe. He wanted to know "for certain" that this would not happen when the RHIC was fired up. The magazine printed this letter, along with a reply by Nobel Prize–winning physicist Frank Wilczek, who stated that physicists are very leery of using the word *impossible* (that is, "for certain"), but that the whole idea of a black hole coming from the RHIC and gobbling up the planet was an "incredible scenario." Here Wilczek used the term "incredible" in its literal meaning: incredible = simply not believable.

Never to be left behind when it comes to sensationalism, the general news media jumped on this possibility with relish. One reporter called the RHIC the Doomsday Machine and said that a physicist told him that its construction was "the most dangerous event in human history." Another account claimed that a grade-school student in Manhattan wrote a protest letter to Brookhaven officials, saying she was "literally crying" as she wrote the letter. The machine was even blamed for creating a black hole that swallowed the airplane that crashed in 1999 killing John F. Kennedy Jr.

This expression of public concern over what might come out of a particle accelerator began some years earlier when Paul Dixon, a psychologist at the University of Hawaii, picketed Fermilab outside Chicago because he feared its Tevatron collider might trigger a collapse in the quantum vacuum that could "blow the whole universe to smithereens." (I wonder if Dixon knows Walter Wagner, as there seems to be something in the air of Hawaii that brings out these kinds of protests!)

This RHIC brouhaha came and went as the machine was put into service in the summer of 2000 with no reported difficulties, at least not any involving the mysterious disappearance of aircraft or untoward corners of the earth being swallowed up in a cosmic vacuum cleaner. But that's not the end of the public's fear of physicists.

In 1994, work began at the European Nuclear Research Center (CERN) near Geneva, Switzerland, on an even more powerful particle accelerator, the LHC that I've already spoken of. After several start-up glitches, the machine went into full service in late 2009, although it won't be ready to operate at its full design level until 2014. This project was the culmination of an idea that had been bandied about at CERN since the late 1980s. And just what was this idea? Nothing less than to build a Big Bang machine that could re-create those fleeting moments nearly fourteen billion years ago when the fundamental building blocks of the universe were put in place.

The engineers at CERN knew that to create the energies necessary to answer questions about the Higgs particle, dark matter, and the like, they had to build a machine more complex than any machine ever created by human beings. In this machine, beams of protons would be accelerated to 99.9999999 percent of the speed of light in an environment colder than interstellar space. The proton beams would then be crashed together in the hope that the particles created in these explosions would yield answers to the foregoing questions.

Another part of the conjuring trick is to actually see the "answers," since the elementary particles created in the colliding beams would decay and disappear in less than a trillionth of a second. To capture these transient objects requires a detector larger than a five-story building, yet so precise it can pinpoint an object with an accuracy of one-twentieth the width of a human hair!

The design and construction of such an incredible device took over ten thousand scientists and engineers more than fourteen years, and an expenditure of more than six billion euros (more than eight billion US dollars), to build.

Just a couple of months before the LHC was supposed to "go live," the same Walter Wagner who expressed concerns about the RHIC filed a lawsuit in Hawaii's US District Court, calling for the US Department of Energy, Fermilab, the National Science Foundation, and

CERN to slow down the LHC preparations for several months in order to reassess the collider's safety. The suit asked for a temporary restraining order on implementation of the LHC and called upon the US government to carry out a full safety study of the machine, including a renewed consideration of the doomsday scenario.

More specifically, the scenario outlined in Wagner's lawsuit includes the following possibilities:

Runaway Black Holes: Millions of microscopic black holes would be created that would persist and somehow coalesce into a gravitational mass that would consume other matter and ultimately swallow up the planet. Most physicists believe that these black holes, if they're created at all, would have minuscule energy and quickly evaporate, thus posing no threat whatsoever.

Strangelets: According to current wisdom, all known matter is composed of various types of elementary objects called "quarks." Wagner and others fear that smashing protons together at enormous energies might create new combinations of quarks, including a nasty version known as a stable, negatively charged "strangelet" that could turn everything it touches into strangelets as well. This is reminiscent of the so-called ice-nine scenario from Kurt Vonnegut's novel *Cat's Cradle* that I briefly described in Part I. Recall that Vonnegut imagined a strange form of matter, ice-nine, that was seeded into the oceans and immediately turned all normal water into a solid crystalline form.

Magnetic Monopoles: All magnetic objects that we know about have two poles, one pointing north, the other south. It has been suggested that high-energy collisions of the LHC variety might give rise to massive particles that have only a single pole, north or south—but not both. The fear is that such a particle might then start a runaway reaction that would convert other atoms into monopole form.

Quantum Vacuum Collapse: Quantum theory postulates

that the vacuum between particles is in fact just brimming over with energy. Some argue that putting enough energy in one place could be enough to break down the forces that stabilize the quantum vacuum energy and allow its release. Calculations suggest that if this were to happen, an infinite amount of energy would be released, creating a massive explosion that would sweep across the universe at the speed of light. The more imaginative versions of this scenario even suggest that perhaps some of the huge explosions observed in other parts of the galaxy might be due to experiments performed by aliens with the quantum vacuum that have gone awry.

So what does the world of science put forth as arguments against these scenarios? Are any of them even faintly plausible? Do any carry a likelihood great enough to trump the innate curiosity of human beings about the universe around us? Let's see some of the counterarguments the world's physics community marshals against these imaginative, if perhaps fanciful, scenarios.

"SCIENTIFIC" FICTIONS

ACCORDING TO EINSTEIN'S FAMOUS EQUATION $E = mc^2$, IF YOU PUT enough mass into a small enough space you could generate a black hole, a region of space with a gravitational field so great that nothing, not even light, can escape from it. Since the LHC will be smashing protons together at near light speed, and protons are made of many smaller particles, it's not totally beyond reason to wonder if several of these pieces might find themselves crushed together into a small enough space to generate a black hole. Here are some reasons why that is very unlikely.

Extra Dimensions: Those worrying about the possibility of mini black holes being created by the LHC assume the energy required to do this is vastly less than what we think is actually required based on studies of the world as we find it. So the possibility of generating a black hole in the LHC arises only in theories that postulate "large extra dimensions." Only in this way is there "room" in these extra dimensions for the type of interactions needed to build black holes at low energies.

Basically, the problem is that producing black holes requires an enormously strong gravitational attraction. But gravity is by far the weakest of the four known forces. So to remedy this difficulty, some theories assume extra spatial dimensions accessible to the carrier of the gravitational force, the graviton, but not accessible to other particles such as quarks, photons, and electrons.

If such extra dimensions really existed, then gravity might actually be a very strong force but still appear weak to us, since the gravitons would spend most of their time in the extra space and seldom visit our part of the universe. At present, though, there is no evidence for the reality of these extra dimensions.

Theory Versus Reality: Strictly speaking, no one has ever seen an actual black hole; it is simply a theoretical construct. Many objects have been observed in the galaxy that are candidates for being black holes. But, in fact, there are a lot of difficulties with the whole notion of a black hole, and we don't really know for sure whether or not they actually exist.

An especially troubling aspect of the whole idea of a black hole is that the general theory of relativity states that time should slow down as an object approaches a heavy object like a black hole. This means that chunks of matter disappearing into a black hole should take an infinite amount of time to vanish, at least as seen from the perspective of an observer outside the so-called event horizon of the black hole. Such an observer would see an object, say a football, floating toward the black hole and then somehow just get "stuck" like a fly on a piece of flypaper at the event horizon. Of course, if you're a quarterback with your hand on that football, you continue on through the event horizon, blithely unaware that anything special has occurred—until you turn around and try to go back out. Then you discover that it's a one-way trip and you've crossed the point of no return. But an observer on the outside would see no such thing; that observer would see you stuck to the event horizon forever.

Cosmic Rays: As early as 1983, Sir Martin Rees of Cambridge University and Piet Hut of the Institute for Advanced Study in Princeton pointed out that cosmic rays have been smashing into things all over the universe for eons. Many of these collisions are at energy levels millions of times greater than the LHC can generate. Yet no planet-sucking black hole has appeared, and the universe is still here. As the world's leading expert on strangelets, Robert Jaffe of MIT, says: "If it were possible for an accelerator to create such a doomsday object, a cosmic ray would have done it long ago." He goes on to state, "We believe there are relevant cosmic ray 'experiments' for every known threat."

So it would appear that until we have accelerators more powerful than the energy of the most energetic cosmic ray, we're covered.

ADDING IT ALL UP

LOOKING AT THE WORLD OF PHYSICS AS SEEN FROM THE STANDPOINT of a theoretical elementary particle physicist, on the asset side of the experiments under way today in Geneva, Chicago, Long Island, and elsewhere, we see the possibility for validation of one of the many competing models of how the universe is put together or the possibility of having to entirely rethink our view of this literally cosmic question. The liability column, which is what the rest of the world mostly sees, contains the extremely remote, but still nonzero, possibility of destroying the earth.

Either of these possibilities is an X-event. Either confirmation or denial of the current Standard Model of physics would be an X-event within the community of physicists, something rare, with great social impact in that community and definitely surprising (especially if the end result is denial). The other case, destruction of the earth by a strangelet, is an X-event impacting a much broader social community, namely, the entire planet (including the physicists!), and would also definitely be a surprise. Of course, I say this partially tongue in cheek as the two X-events are hardly commensurable, the "X-ness" of the second vastly outweighing that of the first. Nevertheless, regardless of how these physics experiments turn out, the end result will without fail be an X-event as we employ that term in this book.

BLOWN AWAY

DESTABILIZATION OF THE NUCLEAR LANDSCAPE

BACK TO THE STONE AGE

IN THE MID-1960S, I FOUND MYSELF WORKING AS A COMPUTER PRO-grammer at the RAND Corporation in Santa Monica, California, while pursuing my postgraduate studies in mathematics at the University of Southern California. By then RAND's heyday as a think tank for the military was already over, as the organization was in the process of reinventing itself as a one-stop-shopping center for various federal, state, and local governmental agencies seeking intellectual salvation for the political and social ills besetting their constituencies. But a few pockets of the "good old days" remained scattered throughout the building, and one day my boss asked me to help out one of those "dinosaurs" who was studying how to best target nuclear weapons. As a result, I ended up programming scenarios addressing the question of how to most effectively bomb Moscow back into the Stone Age. In a touch of irony, just a few years later I ended up living in that very city for the better part of a year as an exchange visitor between the US National Academy of Sciences and the Soviet Academy of Sciences. So I had a chance to actually visit the places I had previously only seen on the map as high-priority strategic targets of opportunity for instant incineration.

Let me give a brief summary of the Cold War mentality that prevailed at RAND and the US defense establishment at that time, and how it has segued into something very different, and in many ways far more dangerous, in regard to the use of nuclear weapons. In the 1960s, we lived in a bipolar world, at least insofar as instantaneous vaporization in a nuclear holocaust goes. The US and the USSR were the only players at the table, with the United Kingdom, France, and China leaning over their shoulders but not really positioned to make unilateral decisions on the use of their much smaller nuclear arsenals. But today there are eight states that have actually detonated nuclear weapons, three more that have conducted nuclear tests, and anywhere from three to seven that are thought either to possess these weapons or did possess them at one time and seem to have given them up in the aftermath of the dissolution of the Soviet Union. And this is not to mention shadowy terrorist groups who are rumored to have purchased a nuclear bomb or two from underground sources. So the situation of who has them and who doesn't gets murkier and murkier as each day goes by. From a game theorist's point of view, let alone that of a military planner or national security adviser, this situation represents a vast "complexification" of the world I was dealing with at RAND in the 1960s.

Today's world of nuclear technology is a case study of complexity overload in action. Not only is there a hard-to-define number of players in the game, the nuclear weapons landscape includes everything from "lost" nuclear weapons from the former Soviet Union, disaffected nuclear scientists shifting their allegiances to the "dark side," continuing efforts by hackers to penetrate the weapons control systems, and nonstate actors like terrorist groups and rogue states on the lookout for a stray weapon or two for sale in the black market. Mix all this together with aging and possibly now unstable warhead systems in even the recognized nuclear arsenals, and you end up with a toxic brew that could result in nuclear weapons going off like firecrackers around the world almost anytime. So the nuclear environment today is indeed a poster child for the way complexity overload threatens to destabilize the entire global power structure—overnight.

A good year to start this story is 1960 with the publication of RAND physicist Herman Kahn's very controversial book, *On Thermonuclear War*. The book took a very dispassionate, objective view of the possibilities and consequences of an all-out nuclear exchange between the United States and the Soviet Union. At the time, the two country's arsenals each numbered around thirty thousand warheads, a vast overkill insofar as wiping each other off the face of the globe goes. By way of comparison, those arsenals have now shrunk to "only" a few thousand each. But the complexity of the overall worldwide "nuclear game" has more than compensated for this decline in raw firepower. Returning to Kahn's book, it's filled with accounts of various types of attacks, the number of people likely to be killed directly or indirectly by lingering radiation, the amount of property destroyed, and the like. Almost immediately upon publication, the book was roundly and soundly denounced by liberal-leaning members of the US Congress for its cold, matter-of-fact manner of addressing what was a highly emotionally charged issue especially then, the instantaneous destruction of many tens of millions of lives. The publicity surrounding the book and its author later led filmmaker Stanley Kubrick to use Kahn as a kind of role model for the Doctor Strangelove character in his film of the same name.

During the first half of the 1960s, questions raised in Kahn's book about how to best gain a strategic advantage in a two-party conflict led to a huge amount of mathematical work at RAND on the theory of games of strategy, a field of study initiated in 1947 by the mathematician John von Neumann and economist Oskar Morgenstern in the context of economic competition. The strategic tension between the United States and the Soviet Union was particularly fertile ground for the development of "game theory," since it involved just two players who presumably acted rationally in choosing their actions at each stage of play. Moreover, it was not unreasonable to assume that the interactions between the players were "zero-sum," meaning that the gain to one was matched by the loss to the other. Such two-person, zero-sum games between rational opponents are actually the only type of game for which a totally satisfactory mathematical theory exists, and for which we can actually calculate the optimal strategies for each player. So while the real Cold War tensions were certainly not a perfect fit to this mold, the approximation was close enough to use game theory to arrive at a number of conclusions that at least seemed reasonable given the idealistic assumptions underlying the theory.

After many years of study, debate, political and military discussion, and negotiation, the principal strategy that emerged for both countries was the now well-known concept MAD, an acronym for Mutual Assured Destruction. MAD began with the recognition that each side had built up a nuclear arsenal (and methods for its deployment) that would guarantee total destruction (many times over) of the other side. So if I'm attacked by the full force of my opponent, I will still be able to completely destroy my adversary in a counterattack. Of course, in order to guarantee that destructive counterattack, a still deadly fraction of my own weapons must be able to survive the initial wave of bombing. This led to the three pillars of the US nuclear attack system—land, sea, and air—consisting of underground missile silos, nuclear submarines, and nuclear-armed aircraft aloft twenty-four hours a day. To the best of public knowledge, this strategy is employed to this very day, despite the fact that MAD's dependability is dramatically reduced precisely because there are now too many players in the game (too much complexity). I'll take up why this is the case at the end of the chapter. As a further point to note, even with just two players the MAD strategy is effective only against a *deliberate* attack by one's opponent.

Unfortunately, a premeditated attack is only one way a nuclear weapon might be detonated. There are many others. In fact, it was argued as long ago as 1958 by Fred Ikle of RAND that the next mushroom cloud is far more likely to arise from a simple accident or miscalculation than from a deliberate, calculated assault. And there are many cases of accidents involving nuclear weapons that could easily have realized this possibility. Luckily, they didn't. Just for the sake of illustration, here are two concrete examples of what might have been.

On January 17, 1966, a collision occurred between a US B-52 nuclear bomber and a KC-135 tanker aircraft while the bomber was being refueled over the village of Palomares in southern Spain. The tanker exploded, causing the B-52 to break apart and scatter wreckage over an area of more than a hundred square miles. One of the four nuclear weapons on board the B-52 landed more or less intact, while the high-explosive "initiators" in two other bombs exploded upon impact with the ground and dispersed radioactive debris over the village and its surroundings. The plane's fourth bomb fell into the sea and was recovered by divers three months after the accident. Not to be outdone by the Americans, here is a similar story from the Russian side.

Common naval practice during the Cold War was for nuclear-powered (and armed!) submarines to shadow the military fleet of the opposing side. During one of these many dangerous encounters of the up-close-and-personal military kind, the American aircraft carrier USS *Kitty Hawk* collided with a Soviet Victor class attack submarine on March 21, 1984, in the Sea of Japan. The *Kitty Hawk* was reputed to have several dozen nuclear weapons on board, while the Soviet submarine was thought to have two nuclear-armed torpedoes. Fortunately, none of these weapons was damaged or lost in the collision.

These are just two examples (of many) illustrating how simple human accidents or miscalculation might set off a serious nuclear incident. These cases show complexity entering into the rising gap between the US and USSR "regulators" (the government and military) trying to rein in the growing complexity of land, air, and sea arms of the nuclear weapons systems. The systems were growing in complexity at a rate far in excess of the regulatory procedures, a gap that was luckily relieved by several small X-events of the sort just described in these examples instead of the "Big One." In fact, during the period 1950 to 1993, the US Navy is known to have had at least 380(!) weapons incidents. During this period, accidents resulted in the loss of 51 nuclear warheads (44 Soviet and 7 US), as well as seven nuclear reactors from lost submarines (5 Soviet and 2 US). A further 19 nuclear reactors from decommissioned submarines have simply been dumped into the sea (18 Soviet and one US).

These figures represent only what is publicly known about US and Soviet accidents up to the end of the Cold War. Given the sensitive nature of the subject, it's a safe bet that much has been kept classified by both sides. Nor is it much of a stretch to imagine that the other nuclear powers have had similar accidents and loss of weapons and reactors. Of course it's one thing to have an accident, even one that spreads radioactivity over a large area. It's quite another to deliberately set off a nuclear explosion. Nuclear weapons have many safeguards built in to prevent just such an accidental detonation from taking place. So far these checks have held, though even without an actual explosion the financial, security, health, and environmental costs of the world's nuclear arsenals is enormous.

NUCLEAR WINTER

AMBIO IS A HIGHLY RESPECTED ENVIRONMENTAL JOURNAL PUBLISHED by the Royal Swedish Academy of Sciences. Around 1980, the editors asked Dutch scientist Paul Crutzen and his American colleague John Birks to prepare a paper addressing the atmospheric effects of nuclear war. As atmospheric chemists, Crutzen and Birks originally intended to look only at increased amounts of ultraviolet radiation reaching the earth's surface as the result of a nuclear war. But by one of those strokes of serendipity that often unaccountably occur in the history of a major scientific breakthrough, they inexplicably decided to consider the smoke from fires as well. Preliminary calculations convinced Crutzen and Birks that there could be enough smoke in a major nuclear exchange to blot out the sun from half the planet for weeks on end. Publication of their paper in the November 1982 issue of *Ambio* stimulated work by many other scientists on the relationship between fire and smoke from nuclear blasts and the darkening of the sun, leading to a major study and meeting in late 1983 that sparked off scientific and public concern over the problem of "nuclear winter."

Subsequent studies indicate that the principal environmental consequences of nuclear war are likely to be: (1) obscuring smoke in the troposphere, (2) obscuring dust in the stratosphere, (3) fallout of radioactive debris, and (4) partial destruction of the ozone layer. This list, incidentally, shows why no such climatic effects were observed during the period of atmospheric testing of nuclear weapons prior to the 1963 Limited Test Ban Treaty. The tests were all conducted over desert scrubland, coral atolls, tundra, and wasteland. These tests set no fires; hence, no smoke.

Let's look at the pieces of this gloomy picture in somewhat more detail.

1. The nuclear explosions immediately send dust, radioactivity, and various gases into the atmosphere. The dust expelled from the surface is enough to build a dam across the English Channel five hundred yards high and thirty yards thick.
2. The explosions ignite fires, burning cities, forests, fuel, and grasslands in the warring countries.
3. The fires send plumes of smoke and gases high into the troposphere. Within a couple of weeks, some of the dust, radioactivity, and smoke is carried around the world by the winds.
4. At the same time, clouds of smoke spread around the earth in the midlatitude zones from Texas to Norway. The dust eventually settles to the ground over a period of weeks to several months.
5. Beneath the clouds of smoke and dust, daylight is reduced to darkness for days and to twilight for weeks.
6. Temperatures drop on the land under the clouds of smoke

and dust. If the nuclear exchange takes place in the spring or summer, this drop in temperature is comparable to the difference between summer and winter ("nuclear winter"). Average temperatures probably do not return to normal for more than a year, and the climate is disturbed for a much longer period.

7. When the dust and smoke clear, the earth's surface is exposed to additional damaging ultraviolet radiation resulting from partial destruction of the ozone layer.

These are the main steps on the road to nuclear winter. Can we really expect such dramatic temperature reductions for such an extended period? Or are these just worst-case scare stories manufactured to direct media attention to a hitherto unappreciated aspect of the horror of nuclear war?

Following publication of the Crutzen-Birks study, Carl Sagan and two of his former students, James B. Pollack and O. Brian Toon from NASA's Ames Research Center, together with Richard Turco and Thomas Ackerman, undertook an extensive set of calculations to check the estimates presented in the *Ambio* paper. The Sagan group had already been sensitized to the possibilities of major climatic disruptions due to dust in the atmosphere by their work on the *Mariner 9* probe to Mars in 1971. It seemed that when the probe arrived, a massive Martian dust storm was under way. While waiting for the storm to abate, Sagan noticed that the instruments on the probe recorded atmospheric temperatures considerably higher, as well as surface temperatures much lower, than normal. Later, Sagan's group began to apply some of the same techniques used in analyzing the Martian dust storm data to similar phenomena generated by volcanic eruptions on Earth. So when the Crutzen-Birks report came out, the NASA team was well positioned to do a detailed computational investigation of the situation.

Using their model, the Sagan group produced a paper that has become famous in nuclear winter circles—and not just for its science. This paper, known under the label "TTAPS" from the last names of its five authors, was published in the prestigious American journal *Science* just before Christmas 1983. To maximize the public exposure of the paper's conclusions, Carl Sagan arranged a prepublication press conference on Halloween to announce the paper's frightening conclusions. As a bit of unsubstantiated scientific gossip, in certain corners of the climatological community it was rumored that Sagan chose this especially dramatic moment to call public attention to the nuclear winter scenario in an effort to garner support for a Nobel Peace Prize nomination. Well, why not? After all, the subsequent East-West dialogue on scientific and political issues surrounding the nuclear winter scenario spawned by the paper's conclusions certainly merits some kind of recognition.

The TTAPS group concluded that a major nuclear exchange in the Northern Hemisphere would result in a short-term temperature drop of almost forty degrees (Fahrenheit) and a total recovery time of nearly one year. By way of comparison, even a one-degree long-term drop in temperature would eliminate all wheat growing in Canada, and a ten-degree drop is typical for an Ice Age. But the TTAPS model was not beyond reproach: it was one-dimensional in its assumption that particles of dust and smoke could move vertically but not spread out in space. This kind of assumption implies that the atmosphere just sits there and radiates energy up and down. In other words, the model allows no movement of energy from one location to another across the surface of the globe or through the atmosphere.

In the years since the pioneering TTAPS study, a number of investigators have substantially extended the scope of the modeling of nuclear blast effects using three-dimensional global circulation models of the type employed by meteorologists to forecast weather. One of the leaders in this effort was the late climatologist Stephen H. Schneider of Stanford University. While Schneider was working at the National Center for Atmospheric Research (NCAR) in Boulder, Colorado, in 1988, he published a summary of the state of nuclear winter modeling.

Schneider and his coworkers concluded that the climatic effects are likely to be much less severe than reported in the TTAPS paper. I hasten to emphasize that this does not imply that there is no problem. Far from it, in fact. However, the projected summertime temperature drops of five to fifteen degrees would be more like the difference between summer and fall than that between summer and winter. This conclusion led Schneider, who shared the 2007 Nobel Peace Prize for his contribution to the work of the Intergovernmental Panel on Climate Change (IPCC), to relabel the overall situation "nuclear autumn." His general conclusion was "It is unlikely that climatic effects of nuclear war alone will be more devastating to the combatant nations than the direct effects of the use of many thousands of nuclear weapons." But the report cautions that the entire assessment hinges upon human behavioral assumptions of what constitutes a plausible scenario for nuclear attacks. More recent studies in 2007 by some of the original TTAPS investigators contradicts this rosy picture, showing that even the smallest nuclear exchange would plunge the earth into temperatures colder than the Little Ice Age (1600–1850), with effects lasting more than a decade. So it appears the threat is very real—and very immediate.

Before I enter further into the horrors of an all-out regional or global nuclear exchange, I'll set the stage by first examining a type of scenario much closer to what one might expect from a 9/11-style terrorist attack involving a "small," 150-kiloton bomb, one about ten times larger than the "Little Boy" bomb that leveled Hiroshima in 1945. This is the type of situation that most concerns terrorism experts, since many nuclear weapons of about this size in the arsenal of the former Soviet Union went missing in the aftermath of the downfall of the Communist regime. And who can say how many more are floating around in the underbelly of the global terrorist movement from weapons facilities in undeclared nuclear powers like Israel or South Africa, not to mention unstable areas such as Pakistan. It's not at all implausible to envision a few of these bombs turning up—or blowing up. To make this picture as realistic as possible, I'll use that poster child for all terrorist attacks, the island of Manhattan, as the setting for this exercise involving a ground-level blast of 150 kilotons, which I noted above is a "small" nuclear weapon.

9/11 REDUX

IT'S A CLEAR SPRING DAY IN MANHATTAN, WITH A SLIGHT BREEZE blowing to the east. People are out at midday enjoying the sunshine and beautiful weather. The usual hoard of tourists cluster around the entrance to the elevator of the Empire State Building, waiting to go up to the observation tower and look out over the city on this perfect day. No one notices the delivery vehicle parked at the curb on Thirty-fourth Street next to the line of tourists. At the stroke of noon, a blinding light bursts from the truck and less than one second later all of midtown Manhattan is simply . . . gone. Even the most heavily re-inforced steel-and-concrete buildings within a half mile of the Empire State Building are totally leveled, as landmark sites like Madison Square Garden, Penn Station, and the New York Public Library just disappear, as if by (black) magic.

The enormous overpressure from the blast doesn't just destroy the buildings, but within the first second kills about seventy-five thousand people, the daytime population of the half-square-mile area surrounding the Empire State Building. Those in the direct line of sight of the blast are completely atomized—no bodies, no ashes, just nothing. Those inside buildings die shortly afterward as the buildings collapse.

Fifteen seconds later, the damage zone from the blast extends for about four miles at an overpressure on its boundary of one pound per square inch (psi). At the leading edge of the outer circle of ruin, which stretches to the Statue of Liberty in the south, into Queens in the east, up to Harlem in the north, and across the Hudson River to New Jersey in the west, there is moderate damage to unreinforced buildings of brick and wood, while reinforced buildings suffer only light surface effects.

Fortunately, a lot of this destruction zone is over water, thus reducing casualties. In the outer ring of damage, there are few fatalities and only about thirty thousand people receive injuries directly from the heat. But the blinding light from the detonation causes permanent retinal burns out to around twenty miles from the center of the blast. Since this is a ground detonation, the eyesight damage is much less than if it had been an air burst several thousand feet above the city.

This discussion has so far ignored the radioactive debris and fallout from the assumed surface blast. The early fallout will be large, much larger than from an air burst. It will slowly drift back to earth creating an elliptical fallout zone concentrated in eastern Manhattan and western Queens and Brooklyn, where within a month 10 to 35 percent of the population will die of radiation exposure.

In summary, the 150-kiloton explosion would destroy over twenty square miles of property, kill more than eight hundred thousand people, injure another nine hundred thousand and lead to additional damage through fires arising from broken gas mains, burning debris, and raw gasoline leaking out of destroyed vehicles (although not so much fire damage as would occur if the blast had been in a more rural area rather than centered in a major city). It's safe to conclude that New York City would never recover its current position as a leading center of finance, culture, and business.

With this chilling story as backdrop, how bad could a regional or global nuclear exchange be? Let's look at a few such "would-be worlds" to get a feel for the possibilities.

MEANWHILE, BACK TO THE TWENTY-FIRST CENTURY

As I write, the international press is reporting a story about Israeli commanders rattling their sabers by calling for a preemptive attack on Iran's nuclear fuel-processing facilities. This particular scenario seems to be the "flavor of the month" right now as the most likely way a so-called limited nuclear war might break out. But as

we'll see in a moment, the very idea of a "limited" nuclear war borders on being an oxymoron, and in the tightly connected geopolitical
structure we live within nowadays, a limited war will almost surely
escalate into something a lot more global than any political leader
would care to admit publicly. Here's a short sketch of how this escalation might unfold.

- As Iran's belligerence escalates against Israel, the Israelis mount
  a preemptive air strike using conventional weapons on Iran's
  nuclear fuel-processing facilities.
- Iran responds with massive rocket attacks armed with conventional explosives, as well as chemical, biological, and radiologi-
  cal warheads.
- Israel responds with nuclear strikes against Iran, and as a preemptive measure against Pakistan, as well.
- Outraged, Pakistan retaliates against Israel, launching a preemptive nuclear strike on their enemy (and Israel's ally) India,
  who then retaliates in kind.
- Israel now attacks Arab and Muslim capital cities, as well as
  "anti-Semitic" Europe and Russia.
- Russian regional commanders open a nuclear assault against
  Israel, its ally the United States, and US European allies. Russia
  attacks China as well, in order to destroy its nuclear capability
  before it has a chance to get out of the silos.
- The United States attacks Russia and also hits China's nuclear
  forces in a preemptive strike.
- China uses whatever nuclear weapons it has left against Russia,
  the United States, and India, which all then launch a counterattack against China.

This scenario, while fanciful in places, is not at all "apocalyptic fiction." And, in fact, anyone who finds the above sequence literally "unbelievable" might look at the sequence of events leading to the outbreak
of World War I as a remedy for narrow shortsightedness and a sluggish

imagination. This is not to say that such a scenario is probable. And, in fact, if it began to unfold, the details would almost surely be different. But those are details; the end result would almost surely be the same: a local event rapidly escalating into a full-scale nuclear exchange involving many, if not all, the declared and undeclared nuclear powers.

The preceding scenario shows the butterfly effect in action, as the seemingly localized, although by no means trivial, shock of Israel bombing Iran's nuclear facilities with conventional weapons quickly escalates into what amounts to a global nuclear exchange. In short, the flapping of small wings in Jerusalem leads to a worldwide firestorm.

The book's Notes and References list numerous sources for scenarios of this kind leading to an all-out nuclear exchange. All these scenarios involve a local conflict (India-Pakistan, China-Taiwan, North Korea–South Korea, Israel–Lebanon/Syria, unknown terrorists anywhere) rapidly spinning out of control, drawing in many nuclear powers and ultimately leading to a global nuclear holocaust. It's worth considering these scenarios within the context of the four principal ways such a scenario might begin.

Aggressive: One or more nations decide to use nuclear weapons against nuclear or nonnuclear nations, in order to promote an economic, political, or military goal. This might occur either as part of an ongoing war or as a first-strike nuclear attack. (The state, of course, may claim it is a preemptive, retaliatory, or even accidental attack.)

Accidental: Since the United States and Russia have "launch on warning" systems that send off rockets before it's confirmed that a nuclear attack is under way, any tensions between these powers can lead to a full-scale nuclear war within thirty minutes of a warning—regardless of how erroneous the warning may turn out to be.

Preemptive: One or more nations believes (correctly or incorrectly) or claims to believe that another nation is developing a nuclear capacity, or if it's already a nuclear power is about to use nuclear weapons against its nuclear, military, industrial, or civilian targets and preemptively attacks that nation. This may result from political or military "brinkmanship."

Retaliatory: A nation or group uses nuclear weapons in response to a nuclear attack—even in retaliation for a conventional explosive, chemical, or biological attack by a nonnuclear nation.

Looking over this list of possibilities, I'm struck by the similarities with such a list that I might have drawn up in the 1960s when working at RAND. But there are also major differences between those more carefree days and the world we live in today. Here are but a few of the differences that make today's world vastly more complex and vastly more dangerous than those halcyon days of half a century ago.

Regional Attacks: Many of the scenarios that appear plausible today involve regional disputes that end up drawing the major powers into confrontation. The Israel-Iran scenario sketched above is a perfect example of this. But there are many others: India-Pakistan or North Korea–South Korea, to name but two. It's difficult to think of any analogous local conflicts from the peak of the Cold War era other than the Cuban missile crisis that had such high potential to escalate into an all-out thermonuclear exchange.

Accidents: I've already noted that accidents have played a big part in nuclear scares since the very inception of nuclear weaponry. But with the collapse of the Soviet Union, along with a dramatically increased number of nuclear power plants capable of going out of control, accidents are more of a factor than ever before in leading to radioactive damage in human populations.

Terrorism: The very mention of a terrorist would have been met with blank stares in the 1960s. Not anymore. The numerous shadowy factions operating around the world, each with its own agenda, taken together with the unaccounted for nuclear weapons from the former Soviet Union and elsewhere, lead to entirely new dangers totally foreign to Cold War–style analyses. And this says nothing about the idea of a "rogue state" like North Korea possessing a nuclear capability. As I explained above, it only takes a single low-yield weapon to bring a major city to its knees—and probably set off a full-scale nuclear exchange in return.

"III" Attacks: The plethora of radioactive materials produced in nuclear reactors at power plants around the world gives rise to greatly increased possibilities for poisoning of populations by so-called Immersion, Ingestion, and Inhalation attacks. The death of former KGB officer Alexander Litvinenko, who drank polonium-210 in a cup of tea in a London hotel in 2007, is a particularly well-chronicled example of this type of attack. Again, very little attention was being paid to such types of catastrophes—accidental or deliberate—in the Cold War era.

One might think that a good way to prevent a nuclear attack would be to take actions to protect populations, target with precision to avoid collateral damage, and the like. Paradoxically, though, it can be argued that most such procedures, while seeming to work against a nuclear attack, actually make such an attack more rather than less likely. Such efforts can actually be seen as actions that widen a complexity gap instead of shrinking it. Before closing this chapter, let's have a quick look at a few of the damage-limitation ideas that have been proposed, indicating why they work to encourage an attack rather than deter it.

Shelters: Growing up as a young lad in West Coast America in the 1950s, I recall stories of people who built underground shelters in their backyards to protect themselves and their families in the event of an "out-of-the-blue" attack by the Soviet Union on the United States. Some of these private shelters still exist but are being used mostly as wine cellars or storage sheds nowadays. But there was considerable discussion at the time about the government building major shelters capable of housing many hundreds, or even thousands, of people. On the surface this sounds like a good idea. But a bit of thought shows it really isn't. Here's why.

First of all, the very presence of a wide network of shelters could easily encourage the leaders of a country to mount a first-strike attack, since their citizens would be protected by the shelters. In fact, the shelters by their very nature serve to remove the "assured destruction" part from the MAD acronym. By the same token, leaders of an opposing country without shelters could well be tempted to launch a preemptive strike since they might feel that they would be attacked. In either case, the psychological climate changes for the worse if one side builds shelters and the other doesn't. Similar arguments apply to any type of space defense systems, such as the Space Defense Initiative (SDI) put forth by US president Ronald Reagan in the 1980s. Of course, if all parties build shelters or deploy SDIs there's no net gain for anyone and the entire situation reverts to the status quo ante. In complexity terms, what's happening here is that the side that builds shelters is increasing its complexity, while the other side's complexity remains fixed. Thus, the shelters widen the gap, making the situation more dangerous than it was before the shelters were built.

Limited War: The concept of a limited war, one that carefully circumscribes the geographical area to be destroyed, has considerable appeal. A so-called surgical strike that takes out an opponent's command headquarters or weapons production facility leaving everything else untouched is often discussed. The scenario outlined a few pages ago involved just such a surgical strike by Israel against Iranian nuclear facilities. But as we saw, it has the potential to immediately become an unlimited rather than "limited" war. Nuclear planners, and even politicians charged with a country's defense, have been asked whether they think full-scale escalation of such a limited attack could be prevented. Their candid reply (once they step away from the microphone) is usually something like, "I haven't the faintest idea!" So it's a huge risk to talk about a limited war thinking that it can really be contained and not break out into a full-scale, global nuclear exchange.

Accurate Missiles: Closely allied to the notion of a limited war is the idea that if we had very accurate missiles that could be targeted to destroy missiles in their underground silos instead of destroying civilians in their beds, this would give rise to a limited war. Development of such missiles leads to the idea of launching a first-strike attack in a "winnable" nuclear war. The underlying argument is that by decapitating your enemy with accurate missiles, their ability to respond is so greatly impaired that "only" a few million people would die in any counterattack.

Arms Reduction: Sometimes people understandably argue that limiting the number of weapons helps reduce the likelihood of a strike, theorizing that fewer weapons corresponds to less danger. After all, each disarmed warhead is one less weapon that could go off. One can also argue just the opposite, namely, that by limiting each side's nuclear arsenal we actually make a first strike more tempting. The basic reasoning is that if a first strike destroys say 90 percent of the opponent's arsenal, then this would leave only a small number for a counterstrike. The game then greatly favors the side that acts first. But if the arsenal under attack is large enough, even the remaining 10 percent is more than adequate to deter the suggested first strike.

ADDING IT ALL UP

GIVEN THE COMBINATION OF AN INCREASING NUMBER OF REGIONAL and local conflicts involving nuclear powers, the vastly increased possibilities of nuclear accidents with so many countries possessing or working toward possessing nuclear weapons, and the continued threat of a terrorist group or groups acquiring "missing" weapons, the overall risk level is very high for a nuclear incident in the relatively near future. Actually, it's quite surprising is that it hasn't happened already. This leads to the unhappy, but inevitable, conclusion that a full-scale nuclear exchange is still one of the gravest threats of all to humankind.

RUNNING ON EMPTY

DRYING UP OF WORLD OIL SUPPLIES

AN ANTI-ANTIPASTO

IN WHAT THE INTERNATIONAL PRESS SAW AS A MINOR BLIP IN THEIR unrelenting search for sensationalist headlines, Italian truckers blocked highways out of the largest cities and border points in Italy in December 2007 in a labor dispute over high gasoline prices and working hours. While the media no doubt saw this strike as just another "pro forma" strike, Italian style (i.e., not to be taken seriously), a look behind the scenes shows some disturbing facts about the fragility of one of the infrastructures we mostly take for granted in the conduct of everyday life.

By the end of the second day of the planned five-day strike, rotting produce was all that remained on supermarket shelves from Milano to Napoli. Fresh meat, milk, fruit, and vegetables had been cleaned out by hoarding consumers, as had many other staples such as flour, sugar, butter, and . . . pasta. Similarly, many gasoline stations had put up signs reading OUT OF FUEL, and long lines of scooters, cars, and vans stood in front of the few stations still operating. Gasoline association spokespersons said that 60 percent of the stations in the country were idle, and that within a day most of those remaining would be closed.

"Since yesterday we haven't had any deliveries," said Ruggero Giannini, sales manager at a supermarket in downtown Rome. "We are powerless in the face of such a situation."

After just two days, an entire country was paralyzed by the failure of a fleet of trucks to deliver the goods everyone just assumes will "be there." Nearly as amazing was the almost total lack of interest on the part of the international press in covering this story. Imagine, a major country on the edge of gridlock in just two days. Yet the antics of pop stars, pumped-up athletes, and politicians push a story like this back to the obituary page. What's really going on here?

Two major messages come across loud and clear from this Italian strike. The first is the "just-in-time" inventory control aspect of the supply chain for commodities like food and fuel that society depends upon for daily life. The second message is even more ominous: the critical role played by oil in the movement of those goods from where they're produced to where they're consumed. The Italian strike shows just how fragile these infrastructures are and the absolutely central role oil plays in enabling the delivery of goods to the population. The fact that the story of this strike was so dramatically underplayed by the media shows how much modern industrialized society has become accustomed to taking the reliable functioning of these infrastructures for granted. People seem to think that baby food, tomatoes, cigarettes, laundry powder, beer, and gasoline just magically appear at the corner store on demand. But what if they don't? That's the multitrillion-dollar question.

In this little antipasto about the Italian truckers, the key element is oil. Oil is the commodity of all commodities. Without it, nothing works in modern society as it's currently structured. So to understand how threatening the truckers' strike really was, we need to take a harder look at how oil supplies might be disrupted not just as a temporary glitch due to angry truckers, dicey weather, or geopolitical maneuvers, but on a long-term basis.

THE SLIPPERY SLOPE

IN 1956, THE AMERICAN PETROLEUM INSTITUTE HELD ITS ANNUAL meeting in San Antonio, Texas. On the program was a talk titled "Nuclear Energy and the Fossils Fuels," to be presented by M. King Hubbert, a geophysicist at the Shell Oil Research Center in Houston. While the attendees at the conference had little inkling of what lay behind this rather bland title, Hubbert's employers certainly did. The head office of Shell was on the phone to Hubbert virtually up to the last minute before he strode to the podium, begging, threatening, and cajoling him to withdraw the presentation. But if nothing else, Hubbert was a stubborn old cuss who stood behind his work, and he just ignored these pleas and unleashed what today has come to be called the "peak oil question," or more informally, "Hubbert's Peak."

What Hubbert argued was that US oil production would peak in the early 1970s, a claim that no one in the oil industry wanted to hear then—or now. His studies showed ultimate reserves of two hundred billion barrels of oil, a claim that got everyone's attention. In fact, actual US production from 1956 to 2000 was a bit greater due to production from Alaska and far-offshore fields in the Gulf Coast. But the difference is small, and the general drift of production is still closely following Hubbert's curve. Today, the whole question of peak oil has been ratcheted up a notch, with many observers claiming that global oil production peaked in about the year 2000. If these predictions are even close to being as accurate as Hubbert's were for US production, the world is in for a major upheaval in every aspect of what we take to be modern life. So it's no trivial matter to understand why Hubbert got it right for the United States, and what connection that type of prediction has for the global society today.

The oil production curve for big oil fields is by now quite well known and understood. It is pretty symmetric, moving upward at around 2 percent a year when the field is new, then declining at about the same rate once it passes its peak production level. So if global oil production peaked a few years ago, what we can expect is an annual decline in production of around 2 percent.

On the other side of the equation is consumption. As world population increases and developing countries in east and south Asia increase their appetite for oil, global demand is expected to move up at a similar 2 percent per year in coming years. Taking both these figures into account, we have an annual shortfall of about 4 percent that must be bridged.

The rational way to solve this problem would be for the market to allocate the available oil to those who need it, with the rich countries subsidizing the poorer ones until alternatives to oil can be found. But the historical record is a bad place to look for rational solutions to global problems (or any other problems, for that matter). Rather, the way to bet seems to be for the occurrence of an X-event of one type or another to step in and release the tension between supply and demand. I'll consider some of these complexity-balancing possibilities a bit later in the chapter.

From a purely geophysical point of view, Hubbert used known principles for how oil is formed, the type of geological regions where it's likely to be found, and other such geophysical properties, together with estimates (some would say "guesses") of consumption rates, known reserves, and the like, in order to forecast that US reserves would pass the halfway mark in the 1970s. For all practical purposes, he turned out to be right. Thus, forecasts that global oil reserves peaked in 2000 have deep implications for how life will look in the second half of this century.

Without any really viable alternative energy supplies waiting to take over for a decline in the availability of oil, we can expect a very different world. A major falloff in long-distance travel, the outbreak of international warfare to secure remaining resources, the disappearance of suburbia, the withering away of globalization, and (not least) the vanishing of the consumer economy are but a few of the likely consequences of turning off the cheap-energy spigot. Of course, these projections rest upon assumptions of their own regarding the likelihood of the discovery of major new oil fields, the development of currently unknown energy technologies, and beliefs about how people will react to sky-high energy prices. I'll return to these matters in a moment. But for now, let's look at the current best estimate for where we stand today in terms of global oil reserves, consumption, and the likelihood of "running on empty" anytime soon.

TANKING UP?!

At a dinner party a few weeks ago, I mentioned the peak oil question to a woman sitting next to me. Let me say right away that this lady was an intelligent, accomplished person with many professional achievements to her credit. So I was a bit surprised when after hearing my "manifesto" about the impending global catastrophe caused by the pump running dry, she remarked, "I thought there was forty years of the stuff left. Are we really running out?" Well, even if there is forty or even fifty or even sixty years of oil left, it's not the running out that's the issue. What really matters is whether we have enough to keep our modern economy going. And that point of failure will come long before the pump runs dry.

Oil geologist and peak oil guru Colin Campbell uses the following analogy, which captures the situation very well. The body of a typical 200-pound man is about 70 percent water. So there are about 140 pounds of water in his body. If he loses even 10 to 15 percent of that water by dehydration, he's in trouble and will suffer severe organ collapse and a lot of other unpleasant things—probably including death. So it's not necessary for the man to lose *all* the water in his system to die. Even a small fraction is enough. The same is true of modern society as it's currently configured. A loss of just a small fraction of the world's daily oil supply is quite sufficient to bring today's industrialized society to its knees. With this thought in mind, let's see where the needle on the oil pressure gauge stands today.

The first thing to understand is that nobody really knows how much oil still remains in the ground. The producer countries all lie like thieves about their reserves for a lot of reasons, both good and bad. Commercial interests and governments collude with the producers in this deception for the usual reasons: money and power. Even with these caveats, there is still a high degree of consensus as to where we stand today.

According to a report in the *Oil and Gas Journal,* at the end of 2005 world oil reserves amounted to 1.2 trillion barrels, of which nearly 60 percent was located in five countries: Saudi Arabia, Iran, Iraq, Kuwait, and the United Arab Emirates. On the other side of the ledger, consumption totaled 84 million barrels per day, with 47 percent of that in another five countries: United States, China, Japan, Russia, and Germany. At present, consumption is increasing at about 2 percent per year. So a billion barrels lasts about twelve days. Doing the arithmetic, this is 30 billion barrels per year. So even if consumption leveled off at today's rate, the 1.2 trillion barrels in reserve would be gone in forty years—just like my dinner companion thought! This is a kind of outer limit. But by the argument sketched earlier of death by dehydration, modern society will go into terminal "petrofication" long before that fortieth year arrives unless the supply-demand ratio is drastically changed.

On the supply side, we have two possibilities: discovery of more oil and/or utilization of alternatives to oil for energy supply. Of course, the first alternative is not really a solution as it simply puts off the day of accounting. At best, it can only buy time to develop the second possibility.

On the demand side, the only game in town is reducing consumption; in other words, a dramatic change of lifestyle from what Western society has become accustomed to enjoying over the last hundred years. I'll return to what such a change might entail in just a moment.

The supply-demand rundown offers a transparently clear picture as to how complexity enters into the peak oil question. We have a continually increasing complexity in society driving the demand, while the complexity of the production side has remained pretty much fixed for decades. The result is that the gap widens and widens. As I've just outlined, the sensible way to narrow it is for both sides to take action to close the gap. But again, a bet on voluntary action to narrow the gap, especially when it involves complexity downsizing on the part of the high-complexity system, is about as likely to pay off as buying a ticket in the Powerball lottery. Downsizing is just not hardwired into basic human nature.

The peak oil question involves a kind of Chinese water torture type of global X-event, one that gradually wears society down to a pale shadow of its former self. But there is the quick death, too. This is what may happen if oil supplies are cut off more or less overnight by somewhat more direct means than just having the tank run dry, most likely some type of geopolitical event or terrorist attack affecting the Middle East. We saw earlier that more than 60 percent of all reserves lie under the sands of a handful of highly unstable countries ringing the Persian Gulf. So it doesn't take much imagination to see how global a small spark somewhere in this part of the world could set off a dire global chain reaction. Let's look briefly at a few scenarios.

SHORT AND NOT SO SWEET

A QUICK GOOGLE SEARCH TURNS UP ALMOST AS MANY SCENARIOS FOR a short-term oil crisis as there are commentators on energy, peak oil, and alternative energy options. Here I've identified four that seem quite plausible and that display the range of possibilities such an oil crisis could take. I hasten to add that these are not predictions or forecasts of what will happen. What will happen is almost surely going to be something different. But like all good scenarios, the ones presented here share a strong family resemblance to what's likely to unfold in the coming years (or days).

Scenario I: Civil War in Saudi Arabia: The two holiest sites of the Sunni Muslim religion, the mosques in Mecca and Medina, are hit by surprise bombings early in the morning. While no one claims responsibility for the bombings, the majority Sunni and Wahhabi population in Saudi immediately assumes the attackers are from the Shi'a Muslim community and launches counterbombings that destroy major Shi'a mosques. Thus begins a bloody civil war that's been brewing for decades.

The Saudi royal family is flown out of the country, as news of the unrest spreads to the capital in Riyadh. Meanwhile, share prices on Wall Street tumble over 5 percent before the exchange is temporarily closed. On the NYMEX commodity exchange, crude oil prices shoot up over $20 per barrel in the space of a few minutes as rumors abound of the blockage of all oil shipments from Saudi.

The rioting in Riyadh spreads across all Saudi Arabia, engulfing the neighboring states of Kuwait, Oman, and the United Arab Emirates. Later, a radical Shi'a group claims responsibility for bombings as Iraq enters into the fray. The Iraqi police and military join the bloodletting. Within days the entire Middle East is aflame, and oil production slows to a trickle.

Scenario II: Nuclear Iran: After Shi'ite elements in Basra declare their independence from Baghdad, Iran seeks protection for its religious brethren by forming an Iran/Shi'ite-Iraq coalition bent on controlling the Persian Gulf oil. The coalition invades Kuwait and Saudi Arabia, seizing the port of Dhahran in the first ten days of a major campaign. Further south, Iran mines the Strait of Hormuz, shutting down shipment of 40 percent of the world's seaborne oil, as well as employing terrorists in boats to block the Suez Canal and mine the Bab el-Mandeb at the southern end of the Red Sea.

Having successfully completed its nuclear arms program, Iran threatens their use if the United States intervenes to defend its Gulf allies. As the United States sets up a ballistic missile defense and air strikes to disable Iran's conventional delivery systems, a nuclear weapon goes off, destroying the major oil shipping terminal of Ras Tanura in Saudi Arabia.

Fearing that Iran might introduce a nuclear device into the United States by covert means, the United States makes a preemptive "limited" nuclear strike intending to destroy Iran's remaining weapons of mass destruction.

With the Strait of Hormuz blocked, the main Saudi oil shipment port of Ras Tanura destroyed, and nuclear weapons going off like Fourth of July firecrackers, this scenario provides plenty of ammunition for a huge disruption in crude oil supply with a correspondingly gigantic upsurge in prices.

An interesting side aspect of this scenario is that it has already been well studied by Pentagon war gamers during the Clinton administration back in the early 1990s. At that time, analysts concluded that even with an Iran having twenty to thirty nuclear weapons and employing an out-of-the-blue attack on its Gulf neighbors, only "irrationality" on the part of the Iranian leadership could provide conditions under which Iran would actually use those nuclear weapons. According to the gaming handbook, "the Iranian leadership in these [nuclear] scenarios would be strongly motivated by religious and nationalist sentiments that might override rational calculations." That was 1992. Today . . . who knows?

Scenario III: Hurricane Houston and al-Qaeda: During the height of the Gulf Coast hurricane season, a massive storm strikes the oil refineries in Texas and Louisiana, crippling the production of gasoline, heating oil, lubricants, and other petroleum products indefinitely. At the same time, al-Qaeda terrorists destroy a large part of the Saudi oil production infrastructure à la Scenario I. In days, global oil prices triple.

Taking advantage of the situation, Venezuela and Iran stir up already brewing crises in their respective parts of the world, putting even further upward pressure on oil prices. A day or two later, simmering conflicts in several parts of the world break out into full-scale warfare in desperate attempts to secure oil at any cost.

Consumers in the United States go into panic-and-hoarding mode as all transportation grinds to a halt, stock markets crash, and riots break out in New York and other major cities as the US economy is ripped to shreds. As just-in-time societal supply lines break down, a massive global depression sets in. Politicians, church leaders, and other "talking heads" haven't the faintest clue as to what is really happening and prove powerless to offer any solution to the hoarding and looting that's occurring around the clock—which only serves to exacerbate the shortages.

This scenario follows the script for one presented on a CNN special in the spring of 2006. And it's no pipe dream conjured up by the overactive imagination of the media, either. Hurricanes do happen. And they happen in just the area where most petroleum products are refined for the US market. It's also a fact that terrorist groups are as opportunistic as anyone else, and the synergy gained by piggybacking on an "attack" by nature may be too good an opportunity to pass up.

It's also a fact that the world economy consists of a dense web of tightly interconnected infrastructures. This web is a very fragile one, and oil is the lifeblood sustaining its operation. When the oil stops flowing, panic can spread throughout the world faster than SARS, bird flu, or any other type of biological epidemic. It's more like an information epidemic, infecting billions worldwide within a couple of days.

Scenario IV: A Supply-Side Coalition: It's May 2014. Oil prices are over $100 per barrel, as both Iran and Venezuela have cut exports by more than seven hundred thousand barrels to punish the developed countries of the West for imposing sanctions. Meanwhile, the US military is preparing to move its

entire Pacific fleet to the Persian Gulf region to combat threats
to the oil fields of the Middle East.

Suddenly reports arrive from Baku of sabotage in Azer-
baijan shutting down the oil fields there, removing a million
barrels a day from the worldwide supply line. Oil prices immediately shoot to over $115 per barrel, stock markets go into
free fall, and confusion reigns supreme in Washington as politicians grapple with the situation.

The US secretary of energy tells his president that supplies
can be released from the Strategic Energy Reserve to reduce
pressure on gasoline supplies. The president ponders this possibility along with the alternative of enforced conservation by
reduction of speed limits and other measures to reduce driving. The military argues that the oil in reserve must be kept
for possible action in the Middle East, while congressional
leaders claim that the population will not accept enforced
conservation. Unable to project military power to central Asia,
the US military is forced to adopt a wait-and-see approach to
the situation.

Fast-forward now three months to August 2014. The situation is immeasurably worse. A secret uranium enrichment
plant is discovered in Iran, confirming Iran's ambitions to build
nuclear weapons. The United States and Israel push for even
stricter UN sanctions. In response, Iran and its vassal state,
Venezuela, threaten to cut back oil production, sending prices
skyrocketing to over $150 per barrel.

The US president enters a critical meeting in the White
House Situation Room, seeing no viable alternative to imposing conservation measures. He knows there is no way to soften
the political and economic blow of what looks to be $200 a
barrel oil staring him in the face. Advisers warn that further
sanctions on Iran will have little effect, since high oil prices
and dwindling supplies act only to further encourage cutbacks
in production by producer nations. On the military side of the

street, there appears to be no alternative but to bring the entire Pacific fleet to the Middle East, thereby ceding control of the Pacific to China. As the meeting ends, the president remarks, "We are facing a mortal threat to our way of life here."

This scenario follows one created by the Securing America's Future Energy and the Bipartisan Policy Center in late 2007. It involved several former top presidential advisers with deep knowledge of the workings of national security matters and wise in the ways of Washington political machinations. The outcome of the one-day exercise pointed to the inability of the US military to project its power over several regions of the globe simultaneously, along with the ability of even minor countries to destabilize the world political and especially economic equilibria.

ADDING IT ALL UP

LET'S SUMMARIZE THE SITUATION. WE HAVE A SHORT FUSE AND A long fuse, both burning toward the same end point: the Last Oil Crisis and the consequent extinction of Petroleum Man. The long fuse is the peak oil scenario in which cheap and easy-to-get oil becomes more and more scarce and therefore increasingly expensive. This scenario involves nothing more than geology (a limited natural resource) and human greed (an unsustainable demand). A shorter version taking us back to a medieval way of life adds the spice of a natural catastrophe like a hurricane or volcano and/or a human intervention such as a terrorist strike to move things along. In either case, it's a matter of a decade or two or three before Petroleum Man leaves center stage—kicking and screaming no doubt—but exiting all the same.

Facing these scenarios, the first thing any reasonable person should (and does) ask is, What can be done? What can I/society do to prevent this "extinction" event? The short answer is . . . *nothing*. The inexorable machinery for this event was set in motion at the moment when a gasoline-powered automobile engine won out over its steam-powered competition early in the last century. The coup de grâce was then delivered after World War II in the ill-fated experiment of living the "American Dream": suburbia. Yes, you too can have it all—live in the country, work in the city. The massive expenditure of resources—energy and money—devoted to the development of highways, shopping malls, trucks (SUVs) masquerading as cars, and the like needed to sustain this "dream" will surely be seen as the greatest misallocation of resources in human history and was the final element needed to seal the grim fate we face today.

At the personal level, someone asked me whether I thought now would be a good time to buy a solar-powered home. I replied that George W. Bush, Dick Cheney, and Al Gore all have state-of-the-art solar-powered "off-the-grid" homes. Bush's has been described as an "environmentalist's dream home," while Cheney's is equipped with state-of-the-art energy-conservation devices installed by . . . Al Gore! Do you think they know something you don't?

To return to the question of what to do, there are a lot of small things that each individual can do that mirror uncannily many of the things environmentalists have been advocating for years, ranging from simple self-education on the nature of the problem to reduction of your personal meat consumption (meat is a very energy-intensive form of food). You might also start learning how to perform emergency medical procedures and thinking about how you're going to survive power outages, food and water shortages, economic breakdowns, and in general, collapse of societal infrastructures.

I'll now look just a bit closer at the two fuses—longer-term peak oil and shorter-term accident/natural hazard/terrorist attack in an attempt to get a feeling for the timing of things. First of all, the long fuse.

According to studies carried out by reasonably unbiased researchers (i.e., those not in the employ, directly or indirectly, of the oil industry, OPEC, national or international energy agencies, political action groups, and other such vested interests), there's a pretty clear consensus that non-OPEC production will peak by the middle of the decade, roughly 2015 or so. As for when global oil will peak, that depends entirely on the situation with OPEC.

If OPEC reserves are greater than the consensus forecast from the objective forecasters, then the global peak might be staved off till around 2020–2025; if the reserves in the Middle East are in the neighborhood of the assumptions made in the models, then the peak will be a few years sooner. In any case, it really doesn't matter much since we're talking just a few years. At today's pace of governmental efforts to take any meaningful action other than blah-blah, those few years are meaningless. On balance, then, the explosion from the long fuse is ten to twenty years downstream.

As for the short fuse, it's a question of how short is short? It could be as short as tomorrow, or even today. But not all possible time frames are equally likely, and it's a good bet that the short fuse is measured in just a few years, say two or three. But it could burn down to the end anytime.

It's not without a touch of irony that the range of catastrophes brewing up from the oil crisis bear a striking resemblance to those put forth by the Club of Rome in their *Limits to Growth* study in 1972. At the time, I recall being present at many sessions at the International Institute for Applied Systems Analysis in Austria, where eminent economists, system modelers, demographers, and other scholars uniformly dismissed these forecasts as ill-conceived. To the defense of the Club of Rome, what was rejected was not so much the forecasts themselves, but the methodological basis upon which the club's researchers had arrived at their conclusions.

The key conclusion of the Club of Rome study was that exponential growth of population and continual energy consumption would precipitate a global economic collapse accompanied by mass starvation. The crisis would take the form of resource constraints in things like energy, food, water, and/or pollution of the environment to such a degree as to make the planet uninhabitable. It's worth quoting the following passage from that study:

If the present growth trends in world population, industrialization, pollution, food production and resource depletion continue unchanged, the limits to growth on this planet will be reached sometime within the next one hundred years. The most probable result will be a rather sudden and uncontrollable decline in both population and industrial capacity.

Both fuses are burning brightly, reflecting the accuracy of this forecast. Pity humanity didn't heed the call in 1972 when something might have been done to derail this out-of-control train when it was still a long way from crashing into the station.

I'M SICK OF IT

A GLOBAL PANDEMIC

MONSTERS AT THE DOOR

IN HIS 1947 EXISTENTIALIST NOVEL *THE PLAGUE*, FRENCH NOVELIST Albert Camus paints a gripping picture of medical workers joining together to fight an outbreak of bubonic plague. Set in the Algerian port city of Oran, the characters in the story range across a broad swath of everyday life from doctors to fugitives to clergymen, all forced to address the very "*Camusian*" issue of the human condition and the vagaries of fate. The end result of these separate deliberations seems to be that humans have at best an illusion of control of their destiny, and that ultimately irrationality governs events. *The Plague* is an account of an event that lies so far outside the reasonable expectations of normal experience—of how life should be—that we see it as simply . . . well absurd, as Camus famously described the conflict between what humans nostalgically expect of existence and the realities of our quixotic, unpredictable, unbelievable world. In other words, Camus's plague was an X-event.

The Plague is but one of a huge number of fictional accounts of an epidemic and its effect on the daily lives of a large population. The surface story line of Camus's tale is that thousands of rats begin to die unnoticed by the residents of the city. Soon, a local newspaper reports this seemingly strange phenomenon and a mass hysteria develops in the populace. In a well-meaning, but tragic, effort to quell the hysteria, public officials collect all the dead rats together and burn them, thus creating a catalyst that actually promotes spreading of the plague. Following a lot of political bickering over what actions to take, the town is quarantined, postal service is suspended, and even phone and telegram service is confined to essential messages. How this latter edict contributes to confining the disease is left a mystery, but it certainly contributes to the sense of isolation felt by the townspeople.

As the plague spreads throughout the city, people eventually give up their petty individual concerns and join together to help one another survive the pestilence. Finally, the plague burns itself out and life returns to normal. People take up their previous daily life patterns, and gradually as life becomes routine, the sense of "the absurd" that the plague revealed is paved over. And so it goes.

In Camus's day, it was relatively easy to confine a disease to a local geographic region, as people didn't fly halfway around the globe for a long weekend in the Seychelles or buy food in their local market that began the day on another continent. But in today's world the plague outlined by Camus would almost surely not be confined to the borders of Oran, but quickly spread to continental Europe and from there to Asia and/or North America and/or South Africa and/or . . . My task in this chapter is to look at the possibility for just such an outbreak and the likelihood of its decimating hundreds of millions of people (or more) before it's run its course.

This story of how the plague spread in Algeria provides a lesson in complexity through the way the individual pieces of the story—the actions taken by the various parts of the city administration and the populace—combine to produce entirely unwanted and unplanned "emergent" effects like burning of the rats that actually contributes to spreading of the plague instead of containing it. So the real story of this disease that makes it a complexity-generated X-event is not the outbreak of the disease itself, but the way the human systems interacted so as to exacerbate the death toll rather than reduce it.

Before setting sail on this voyage through the world of viruses, bacteria, and other nasty, dangerous, and infectious things, let me first clarify a bit of terminology that I'll use throughout this chapter.

- *Incidence*: The number of new cases of a disease that appear in a given population of a specified period of time.
- *Epidemic*: An excessive and related incidence of a particular disease above what is normal for a given population. For instance, Camus's plague was an epidemic.
- *Pandemic*: An epidemic that spreads beyond a given continent and becomes a wide-ranging problem. AIDS today is a pandemic.
- *Endemic*: A disease having a relatively low base-level incidence rate, but not necessarily constant. The common cold is the most typical endemic disease in just about any population.

With these definitions at hand, we see that epidemics and even pandemics are far from a new phenomena. They have been with us for just about as long as humankind has walked the planet. And they're not going away anytime soon. Just to put some meat onto this skeletal statement, here is a short list of some of the more infamous and deadly outbreaks of such diseases over the last couple of millennia.

- *The Antonine Plague (AD 165–180)*: A suspected outbreak of smallpox that decimated Rome for more than a decade, killing five thousand people a day at its peak. Estimated number of deaths: five million.
- *The Plague of Justinian (AD 541–750)*: This was probably bubonic plague in the eastern Mediterranean area. The disease began in Egypt and quickly spread to Constantinople and then

to Europe and Asia. According to chroniclers of the period, the disease was killing ten thousand people a day in Constantinople at its peak. Estimated number of deaths: one-quarter to half the human population in the areas where it was active.

- **The Black Death (1300s–1400s and beyond):** A pandemic outbreak of bubonic plague in Europe, the Middle East, China, and India. Estimated number of deaths: one hundred million over a period of two hundred years.

- **Spanish Flu (1918–1919):** The "Great Influenza" was almost surely the deadliest pandemic in history. It is said to have started in Haskell County, Kansas, and was then transmitted though movement of soldiers at the end of the First World War. Estimated number of deaths: one hundred million. In contrast to the Black Death, which killed over a period of centuries, the Spanish flu claimed a similar number of victims in just six months. To put this figure into perspective, given that the world's population is now about four times larger than it was in 1918, the same illness with the same level of lethality would now strike down over 350 million people worldwide.

- **AIDS (1981–present):** Most likely this is a virus that "jumped" species from monkeys to humans in Africa a few decades ago. Estimated number of deaths: twenty-five million and climbing.

This chronicle could be greatly extended but the point is clear. Epidemics and their much nastier relatives, pandemics, richly deserve their position as one of the Four Horsemen of the Apocalypse. But the foregoing list is just a summary.

One might wonder where these killer diseases come from and whether they have been around since living organisms crawled out of the primeval soup. According to recent work by Nathan Wolfe, Claire Dunavan, and Jared Diamond, major human diseases are of fairly recent origin. For the most part, they have arisen only after the development of agriculture. This work identifies several different stages through which a pathogen that originally infects only animals can evolve into one that exclusively infects humans. The main point in this research for us is that diseases leading to epidemics can arise from sources that originally have nothing to do with humans, at all.

To understand the likelihood of another killer plague, we need more information about not just how these infections start, but also about how they spread through a population. To this end, let's look at how a modern plague, Ebola fever, has unfolded over the course of the last quarter century or so.

SAME STORY, NEW CAST

IN 1976, MABAKO LOKELA WAS A FORTY-FOUR-YEAR-OLD SCHOOL-teacher in Zaire. Returning from a trip to the north of the country in late summer that year, he became sick with a very high fever. During the next week, he started vomiting and began to bleed from his nose, mouth, and anus. He died less than a week later. At the time no one could pinpoint the cause of his death, although he is seen today as the first victim of what we now call Ebola fever.

Not long after Lokela's death, more than three hundred other patients began turning up with the same symptoms. The overwhelming majority of them died within a couple of weeks. Thus, Ebola fever came onto the radar of the international medical community as perhaps the most virulent disease ever to infect humans.

Thirty years after the first outbreak, the precise origin of Ebola is still unclear, although some evidence points to fruit bats as the carrier. What is known is that the disease migrated somehow from the African jungle to the outskirts of Washington, D.C., in 1989, and a secret military SWAT team of soldiers and scientists was mobilized to stop the virus breaking out in the nation's capital.

What does it take for a pathogen like Ebola to spread through a population? And what are the warning signs we should be looking for as a signal of an epidemic in the making?

The first point to note is that when it comes to infectious diseases,

not everyone is created equal. Some people are simply better positioned genetically and socially to transmit the disease than others,
their immune systems having the ability to tolerate the disease in its
infectious stage long enough to pass it on before either succumbing
to or recovering from the infection. In severe acute respiratory syndrome (SARS), a Chinese physician spread the infection to a number
of people in a hotel, who in turn took the outbreak to other Asian
countries. The disease ultimately spread to more than thirty countries
around the world and killed over eight hundred people.

Epidemics are a function of the disease pathogen itself (the virus or
bacteria), the people who actually have the disease, and the connective
structure of the overall population in which the infected people circu-
late (the interaction patterns of infected and uninfected people). This
process has a striking parallel to the spread of information throughout
a population, in which an idea spreads from one person's brain to that
of another person instead of it being a virus or bacteria moving from
body to body. Formally, the two processes are identical, other than
that in one case the infectious agent may be a few bars of a popular
song or a computer virus, while in the other it is an infectious biological agent.

Best-selling writer Malcolm Gladwell has described the process
of the outbreak of an information epidemic in his book *The Tipping
Point,* where he identifies three laws of epidemics: the Law of the Few,
the Stickiness Factor, and the Power of Context. These so-called laws
parallel similar principles used by epidemiologists to characterize and
model the spread of a disease through a population. So let me briefly
summarize each of them:

The Law of the Few: There exist "exceptional" people in a
population who are extremely well connected and at the same
time strongly virulent. As a result, these few special people have
the ability to expose a disproportionately large number of the
population to the infectious agent. In the lingo of the epide-
miological community, such people are termed "superspread-

ers." An outbreak of SARS in Toronto, for instance, was traced to such a superspreader.

The Stickiness Factor: This law says that relatively simple changes can be made to many pathogens enabling them to "hang on" in a population, year after year. Influenza is a good example, where each autumn new strains of last year's virus appear, each a slight modification of what came before, the changes being just enough to enable the virus to slip through the immune systems of many people and infect a large fraction of the population.

The Power of Context: This law asserts that humans are a lot more sensitive to their environment than it may seem at first glance. In other words, whether people are ready to change their behavior and, for instance, voluntarily quarantine themselves or even take basic precautionary measures to avoid infection, such as wearing a mask or washing their hands, depends on the cultural standards of the particular population they are a member of. In a small town, people will react differently than in a major metropolis. And that difference may be the difference that matters insofar as whether an epidemic breaks out or not. This is a good point to interject a few words about where our ideas of complexity enter the pandemic story.

At the level of a growing complexity gap leading to an X-event, the picture is rather clear, at least for an individual. We have two systems in interaction, the pathogen and the human immune system. Each has its own complexity level, determined in one case by the tools that the pathogen can employ to penetrate the immune system's defenses, pitted against the tools the immune system can bring to bear to resist the attack. As long as these two levels of complexity stay more or less in balance, no infection takes place. But when the pathogen mutates faster than the immune system can react, that's when trouble begins. And as that gap between the two systems widens throughout a major fraction of a population, an explosive level of infection can occur. Eventually the gap is narrowed as the immune systems of the population finally adapt to the pathogen. But the speed of complexity increase on the two sides of this "arms race" may be very different, accounting for the many years it often takes for a plague-style pandemic to run its course. This arms-race type of complexity gap is at the level of individuals. But there is also a population-level complexity story, as well.

The three network principles outlined above by which infectives interact with those who are not infected and pass a virus or bacteria to them is a network complexity issue. In particular, studies by network analysts like Duncan Watts and Albert-László Barabasi have shown that there are critical levels of connectivity in the linkages among the population at which an infection can suddenly "take off" like a wild-fire. The threshold between containment of the disease and its going "viral" (that is, literally out of control) is a very fine line, an illustration of the butterfly effect I discussed in Part I.

So these are the rules of the game by which an epidemic of either a disease or a rumor breaks out and spreads. What are the stages we should be aware of that give us an early-warning sign of an epidemic in the making?

According to the World Health Organization (WHO), an influenza pandemic has six distinct phases, ranging from the appearance of an influenza virus subtype in animals with low risk of human infection in Phase 1, to sustained transmission of the virus in the general human population in Phase 6. The various phases constitute an increasingly clear set of signals, or "fingerprints," that a pandemic is brewing. Here is a summary of all six phases:

Phase 1: No new influenza virus subtypes have been detected, but an influenza virus subtype that has caused human infection may be present in animals. If present only in animals, the risk of human infection or disease is considered to be low.
Phase 2: No new influenza virus subtypes have been detected in humans. However, a circulating animal influenza virus subtype poses a substantial risk of human disease.

Phase 3: Human infection(s) with a new subtype, but no human-to-human spread, or at most rare instances of spread in a close contact.

Phase 4: Small cluster(s) with limited human-to-human transmission but the spread is highly localized, suggesting that the virus is not well adapted to humans.

Phase 5: Larger cluster(s) but human-to-human spread is still localized, suggesting that the virus is becoming increasingly better adapted to humans but may not yet be fully transmissible (substantial pandemic risk).

Phase 6: Increased and sustained transmission in the general population.

As an example of the use of this list to characterize the stage of a possible pandemic, the so-called bird or avian flu, technically labeled the H5N1 virus, is currently at Phase 3. A movement up to Phase 4 would represent a huge increase in the danger to humans, as it is the first phase at which human-to-human transmission would be confirmed. This would no longer leave the virus in the careful monitoring stage, but enormously elevate the importance of seeking a vaccine and initiation of preventive public health measures.

PUBLIC HEALTH, PRIVATE LIVES

As recent legislation has shown, health is no longer a private matter. For instance, due to the health hazards of inhaling even secondhand cigarette smoke, the United States and many nations in the European Union (but alas, not Austria) have banned smoking in all public places, including restaurants, bars, and cafés, on the grounds of health. Bear in mind here that secondhand cigarette smoke doesn't have nearly the infectious or immediate life-threatening potential of something like Ebola fever, Spanish flu, or even tuberculosis. So where do we draw the line between the limitation of personal freedoms and public health?

A particularly graphic illustration of this dilemma occurred in the early 1900s with the cook Mary Mallon, known to history as "Typhoid Mary." She was an immigrant from Ireland who worked in the New York City area between 1900 and 1907. During this period, she infected more than two dozen people with typhoid fever, even though she showed no signs of the disease herself.

People catch typhoid fever after eating food or drinking water that has been contaminated through handling by a human who is a carrier of the disease. Mary Mallon almost surely had had a bout of typhoid herself at some point in her life, but the bacteria survived in her system without causing further symptoms.

When public health authorities confronted Mary with the news that she was a possible carrier of the disease, she strongly denied several requests for urine and stool samples. Part of her argument was that a local chemist had tested her and found she had no signs of the disease-causing bacteria, at least at the time of the testing. Eventually, the New York City Health Department placed her in quarantine, isolating her in a hospital on North Brother Island for three years. She was released under the condition that she would no longer work in the preparation and serving of food.

But Mary was having none of this; she adopted the pseudonym "Mary Brown" and returned to work as a cook. In 1915, she infected twenty-five people at Sloan Hospital in New York, after which she was again taken into custody by the health authorities and returned to quarantine where she spent the remainder of her life. Typhoid Mary died in 1938—of pneumonia, not typhoid—and was cremated.

The case of Typhoid Mary illustrates perfectly the ethical dilemma facing public health officials: How do they "properly" balance the rights of Mary Mallon to freedom of movement and employment with the rights of the public to be protected against life-threatening actions and behaviors by other people in society? This is the *intrasociety* dilemma. There is an *extrasociety* version, as well: How does a country balance the right of movement of people across its borders with the right to protect their nation from emerging infections? Let's look a bit more at these two situations.

In late 2006, the WHO announced an outbreak of tuberculosis (TB) in the KwaZulu-Natal region of South Africa. Alarmingly, of the 544 patients in the WHO study, nearly 10 percent had a new strain of TB that was resistant to not only the so-called frontline drugs, but also to at least three of the six "backup" treatments. The median survival time for these *extensively drug-resistant TB (XDR-TB)* cases was just sixteen days.

Coupled with the high level of HIV infections in the country, South Africa also suffers from a huge level of infectives who fail to comply with the drug regimes they're given to cure the TB. WHO estimates that 15 percent of patients fail to complete the frontline programs, and a staggering 30 percent default on backup drugs. This has led to an overall cure rate of only about half the patients, which has made XDR-TB not only a potential national disaster in South Africa, but threatens the world at large, via South Africa's steadily increasing tourist population.

To stem the spread of XDR-TB, a number of very severe social measures have been proposed, ranging from restoring social welfare benefits to hospitalizing patients so as to encourage them to remain hospitalized to far more extreme measures such as forcibly detaining people with XDR-TB. The WHO currently recommends that such patients voluntarily stop mixing with the uninfected population. But there are no measures to enforce this separation. The South African government has thus far not been willing to employ detention as a public health measure. All this despite the fact that international law does allow for precisely this type of forcible restraint if all other measures to stop the spread of disease have failed.

So the situation in South Africa with respect to XDR-TB is a living example of the conflict between the removal of an individual's freedom of motion and right of assembly so as to shield the general population from a killer disease.

Here is another huge threat looming on the horizon.

Chinese hospitals derive a substantial fraction of their income from selling drugs to patients. As a result, doctors routinely prescribe multiple doses of antibiotics for routine problems like sore throats. This has led to a dramatic increase in the evolution of antibiotic-resistant strains of bacteria.

Warnings are already being given about the spread of these new strains through international air travel and food distribution, as Chinese pigs imported into Hong Kong in 2009 already showed signs of being infected with these "superbugs."

As super-resistant strains of bacteria start appearing at ports of entry around the world, nations are starting to face ethical conundrums when it comes to sealing their borders against travelers and immigrants possibly carrying a communicable disease. Is there anything a country can do to protect itself from this type of threat?

During the SARS outbreak, the government of Singapore installed thermal-imaging scanners at all ports of entry into the country—sea, land, and air. The body temperature of everyone entering the country was checked prior to their going through immigration to see if they had a fever. This was a simple, nonintrusive screening procedure, comparable as a nuisance factor to normal airport security checks. But we can't say the same for other possible measures to control the importation of a disease at a national border.

In the United Kingdom, there have been calls for compulsory screening of all immigrants for TB and HIV. While debatable as an effective measure for keeping these diseases from crossing the border, there is no debate whatsoever on the practical and ethical questions such a procedure raises. For example, which immigrants are singled out for screening? All of them? Just those from certain countries? Only asylum seekers?

We see that such border "filters" give rise to the possibility of discrimination, loss of privacy, and a certain kind of stigma. To put it compactly, stopping diseases at the border is not quite the same thing as controlling immigration.

So what can we do realistically to prevent a pandemic?

There are at least three ways to stop a pandemic:

Eliminate Infected Animals: By slaughtering the entire poultry population of one and a half million birds, the authorities in Hong Kong stopped the avian flu virus H5N1 in its tracks after the initial cases of human infection were reported in 1997. Unfortunately, this process was both hugely expensive and not totally effective, since the virus has reappeared since then. Nevertheless, the procedure has some measurable effect, at least if the virus can be localized. Similar mass culling was used in the United Kingdom in 2001 to stem foot-and-mouth disease in cattle, when four million animals were slaughtered. But this "shotgun" approach to stopping disease raises many troubling questions, not the least of which is who is going to compensate the farmers for the loss of their animals, hence their livelihood?

Vaccination: Protecting animals and humans by vaccination is also a tricky business. For instance, even if a vaccine exists, it may be impractical to administer it to large numbers of people or animals. Moreover, it's difficult to distinguish a vaccinated animal or human from one that's not. So restriction of movement of the nonvaccinated may be difficult to monitor or control.

Drugs: In contrast to vaccines, which are preventive measures, drugs are an after-the-fact treatment to prevent the outbreak of a pandemic. For bacterial infections, there now exist many very effective antibiotics. There also exist a growing number of bacterial strains resistant to such drugs, like the XDR-TB noted earlier.

When it comes to viruses, the situation is far worse. The only effective antiviral agent for the avian flu virus H5N1 seems to be Tamiflu, which acts as both a kind of vaccine in that it prevents infection, as well as a drug that enhances the survival rate of those already infected. But in either case, it must be given very shortly after exposure to the virus or contraction of the infection. Moreover, variants of the virus have already turned up that are resistant to the normal Tamiflu treatment. So yet again there exist no magic bullets for all known infective agents.

Strangely, perhaps, the most effective general procedure for stopping an outbreak from becoming a full-fledged pandemic is simple common sense. The key element is educating the population about elementary procedures for health care and sanitation. For example, washing your hands when handling food, keeping your home and outdoor areas clean, properly taking medications, and other such procedures go a very long way toward stopping infectious diseases before they can develop into a pandemic or even into an epidemic.

But what about stopping pandemics before they have a chance to get off the ground? Do we have any procedures for effectively forecasting the outbreak of something like avian flu or SARS? This takes us into the realm of how to model the way an epidemic or pandemic takes place once an infection has gotten a foothold in a population. So let's look at some unexpected directions that researchers are taking in understanding the way diseases spread both in space and in time.

PLAGUING PATTERNS

VIRTUAL PLAYGROUNDS, LIKE THE ENORMOUSLY POPULAR MULTI-player online games *World of Warcraft* or *Second Life,* have a following that numbers in the hundreds of thousands. In *World of Warcraft*, players interact in real time on the Internet using computer-controlled avatars to fight battles, form alliances, and gain control of territory.

At first glance, *World of Warcraft* hardly seems like a testing ground for real-life battles against influenza, SARS, bubonic plague, or any other type of communicable disease. But first impressions can be deceiving. And work in the United States by Nina Fefferman of Rutgers University and her collaborator, Eric Lofgren of Tufts University, is showing how these virtual worlds can provide deep insight into the way pandemics form in the world we actually inhabit.

For many decades, mathematical epidemiologists have been creating models of the spread of disease, in order to try to understand and predict the outbreak and spread of epidemics. Unfortunately, to make these models mathematically tractable it's necessary to introduce a host of simplifying assumptions that sometimes assume away the very questions they're trying to answer. So computer games, which allow an almost unlimited variety of detailed behavior by the players to be incorporated into their actions, seem to be a good way to overcome some of the limitations of the math, say Fefferman and Lofgren.

The collaboration between the two scientists and the game's publisher Blizzard came about when programmers introduced a highly contagious disease into a newly created zone in the game's hugely complicated environment. Initially, the "patch," as such new elements are termed, worked as envisioned: veteran gamers recovered from the disease, while inexperienced players succumbed and were left with severely disabled avatars.

But soon things began to go out of control. Just as we see in the real world, some of the infected avatars made their way into the heavily populated cities in the virtual world and infected the inhabitants. The disease also spread through infected domesticated animals, who were quickly abandoned by their owners and left to roam aimlessly through the world infecting other animals and avatars. In short, it was a virtual pandemic.

Programmers at Blizzard tried setting up quarantine zones. But in the virtual world, as in the real world, quarantines were ignored as avatars tried to escape so as to carry on their battles. Finally, the programmers had to shut down the servers and reboot the system in order to eliminate the disease and make the game playable again. Roll back the system! Wouldn't it be nice to be able to do that in reality.

Lofgren was actually playing the *World of Warcraft* when the plague struck, and he immediately saw the potential for using the game as a testing ground for studying the spread of disease. What intrigued the researchers was the opportunity to study how people *really* behave in a public crisis as opposed to the behavioral assumptions made in the earlier mathematical models. People are very different from the homogeneous agents that populate the worlds of the mathematical epidemiologists. In those mathematical models, the individuals in the population all had the same properties regarding virulence of their infection, ability to infect others, and so forth. The heterogeneity possible in the computer models can make a huge difference as to whether a disease breaks out into an epidemic or not, argued the researchers. How many will try to escape a quarantine? How many will start to cooperate if they are scared, as in Camus's story? As Fefferman says, "We simply don't know."

This is where the virtual worlds comes into play, since the players can be given individual characteristics for virulence, resistance to infection, cooperation, escape, and so on. The system can then be "turned on" to see what happens. Skeptics argue that players might be much more ready to take risks in the virtual world than in the real one. The counterargument is that players have invested considerable amounts of time and energy into making their avatars strong and in forming alliances. As a result, a lot of the players' egos are invested in their virtual representative, and they don't want to see their egos crushed by taking outrageous risks.

In the end, of course, the virtual world simulation is just that: a simulation. Like any model, it is not a perfect mirror of the real world. There are assumptions built into the virtual world, too. Still, it seems to be a promising step toward understanding how potential pandemics spread and, most important, how they can be stopped before they have a chance to get off the ground.

ADDING IT ALL UP

BEFORE SUMMARIZING WHAT WE'VE DISCOVERED ABOUT PANDEMICS, I'll address very briefly a point that appears regularly in the popular press and elsewhere regarding pandemics: the question of bioterrorism.

Everyone can agree that bioterrorism is a potential problem. No doubt about that. And it's a problem that perhaps deserves even more attention, or at least more resources, than it currently receives from governments around the world. But from the standpoint of my goals in this chapter, it doesn't really matter much whether a pandemic arises from accidental or intentional human actions. The dynamics of the spread of disease and the end result are indistinguishable. For that reason, I have said nothing in this chapter about detection, prevention, and/or mitigation of terrorist attacks using biological weaponry. Now back to our story.

We have seen that even without the benefit of terrorists, nature is perfectly capable of brewing up a vast array of biological threats to human existence. Epidemics and pandemics of a bewildering variety have regularly made their appearance on the historical stage and can certainly be expected to reappear in various guises again. That goes almost without saying. The real question is whether humankind will be prepared to deal with a major pandemic when and if it occurs.

As to that question, the outbreak of a worldwide life-threatening disease can happen anytime. In fact, sooner is more likely than later for the twin reasons of the worldwide trend toward migration to the cities, thus leading to greater urban population densities, together with the rather lackluster international cooperation on the formation of monitoring and prevention of disease. People simply don't want to take seriously another outbreak of Spanish flu or SARS or avian flu or . . . . But they are "out there." And they will get you—if you don't watch out!

DARK AND DRY

FAILURE OF THE ELECTRIC POWER GRID AND CLEAN WATER SUPPLY

I. DARK

NO POWER TO THE PEOPLE

July 13, 1977, was a hot, humid night in New York City, when at around 8:30 p.m. the lights suddenly went out—and stayed out for the better part of twenty-four hours. I vividly recall this power failure, as I was a professor at New York University living in Greenwich Village at the time. Unhappily for my wife and me, we were living on the seventeenth floor of a university building at Washington Square. Luckily for me, though, I was in California on a consulting job at the time of this power failure, leaving my wife to carry the burden of the event for our family. For the period of the outage she faced a grueling trip up the stairway lugging bottles of water for cooking, drinking, and bathing, together with food and other necessities of daily life. According to her accounts, experiencing this unelectrified Manhattan was eerily like living in a dream world, as the streets of the Village were turned into the venue for improvised street parties. People strolled out to experience and witness the city without power, and everyone was discussing the situation. But only those with battery-powered radios had access to official information about the event or any idea when the power was likely to be back on.

I learned later that other parts of the city were far less tranquil. Outbreaks of violence, looting, and arson were reported in Harlem, Brooklyn, and the South Bronx. People smashed store windows to get electronics, jewelry, clothing, furniture, and other consumer items, not to mention food. More than a thousand fires burned, at least six times the normal rate for that time of the year, and seventeen hundred false alarms were reported. Though these more troubling and threatening aspects of the power failure were not seen in the Village, it seems likely that just one more day of the outage would have been enough to bring them to the southern part of Manhattan, as well.

It's worth noting that this power failure was strictly a New York City affair, in contrast to the failure twelve years earlier that blacked out the entire Northeast and parts of Canada. In 1977, all five New York boroughs were dark within an hour, along with parts of Westchester County immediately north of the city. As it turned out, the surface cause of the failure was what the power company ConEd called an "act of God." Four lightning strikes, the first at 8:37 P.M., knocked out power lines feeding the city's grid. With each successive strike, neighboring power companies in New Jersey, New England, and Long Island disconnected their grids from New York City, to protect their own systems from damage and to serve their own customers. It's interesting to compare this localized, relatively minor failure with the Great Northeast Blackout of 1965 and the far more recent New York power-grid collapse in 2003, the two biggest blackouts in recorded history.

THE 1965 BLACKOUT

November 9, 1965, was not a day when air conditioners were running full out. Nor was it a time of especially heavy electricity demand, in general. Nevertheless, it was the moment for what became known as the Great Northeast Power Blackout, by all measures the largest electricity failure ever experienced up to that time, extending from north of the Canadian border in Ontario to as far south as New York City and eastward to western New Hampshire and Cape Cod. Thirty million people in eight states and the province of Ontario were affected for differing periods of time. So what happened?

The blackout started in Canada at Ontario Hydro's Beck Power Station near Niagara Falls. At 5:16 P.M., a relay on one of the Toronto-bound transmission lines failed, tripping a circuit breaker and removing the line from service. When it occurred, another station supplying Toronto was also down, and as the demand was high in Toronto for winter lighting and heating, the grids were already operating at close to full capacity. The two failures combined to trip circuit breakers on four other lines, thus shifting the load southward to lines leading to the United States.

The power surge southward tripped connections with the PASNY utility lines, destabilizing the principal transmission paths in New York State. Within seconds, the Canadian grid was decoupled from the New York grid. Seconds later, the destabilized systems caused a cascade of further line failures: New England, downstate New York, and other areas quickly shut down. This cascade of disconnections quickly broke the entire power supply network into disconnected islands within a few seconds of the initial failure. Each island then had either a deficiency to try to make up or an oversupply that had nowhere to go. The imbalance led to further failures, and within a few minutes more than thirty million people were without power. For a variety of reasons, New York City was the area without power for the longest period. But on this occasion New Yorkers proved themselves to be a pretty hardy and adaptable lot, and they suffered the discomfort in a fairly good-natured way—*without* major outbreaks of street violence, looting, or other lawless activity.

In the wake of the blackout, much more effective computer controls were instituted on the entire grid, and the North American Electric Reliability Council (NERC) was established to bring together the various independent operators composing the power-supply network in order to establish operating standards for moving electricity from one region to another.

By the standards of the 1965 blackout, what happened in New York City in 1977 was pretty small potatoes. As already noted, lightning strikes were the proximate cause of the blackout rather than a mechanical failure in the system itself, and the geographic area affected was mostly restricted to New York City and its immediate environs. But the social impact was dramatically different. In the twelve years since the 1965 blackout, the social climate had changed substantially—and not for the better. So by the time the lights went out in New York in 1977, it was not a time for gaiety and block parties but rather an opportunity for looters and other lowlifes to crawl out from under their rocks. The lawlessness damaged the image of New York for many years thereafter. The contrast between the reaction of the populace in New York City to the 1965 and 1977 power blackouts illustrates better than any academic theory could possibly show how much the prevailing "mood" of the population at the time of such an incident determines how the social behavior will be during the period in which the crisis unfolds. Now let's fast-forward to yet one more major power disruption, the biggest blackout of them all.

THE 2003 EAST-MIDWEST BLACKOUT

Just after four o'clock in the afternoon of August 14, 2003, the failure of a power station at FirstEnergy's generating plant in east-central Ohio set off a cascade of failures that spread like a wildfire from the

US Midwest up into Ontario and onward to the Northeast of the United States, ultimately taking out electrical power to more than fifty million people. The cascade of rolling failures unfolded over a period of just eight minutes.

A task force investigating the failure identified a number of causes, mostly pointing the finger at FirstEnergy's violations of different NERC standards: operating its system at inappropriate voltage levels, not recognizing or understanding the deteriorating condition of its system, failing to properly manage tree growth next to its transmission lines, and on and on. In short, basic human error on the part of the management of FirstEnergy was the immediate cause of the blackout.

In the aftermath of the blackout, many called for a complete makeover of the power grid. Everyone recognized that the current system was old and deteriorating at a time when power needs were increasing rapidly each year. Restructuring was desperately needed, and a new grid with built-in reliability long overdue. Here we see a classic complexity overload situation at work. One system was an old, deteriorating, low-complexity power grid infrastructure, saddled with low-tech hardware ranging from coal-fired power generation stations to failing transmission lines and on to out-of-date software trying to manage control systems designed decades ago. This system is pitted against the increasingly complex needs of household consumers powering up a bewildering array of equipment, not to mention firms and institutions calling for satisfaction of their own electrical power requirements. This growing gap is an X-event just waiting to happen— and inevitably it does as the examples just cited illustrate.

As of this writing no substantive actions have yet been taken to address these matters. This is not just a North American problem, either. For the sake of a more global perspective, let's have a quick look at similar situations that have taken place in recent years in other parts of the world.

• • •

THE YEAR 2008 GOT OFF TO AN OMINOUSLY POOR START IN SOUTH Africa as rolling blackouts struck the largest cities two or three times a day from early January onward. Initially, these blackouts seemed little more than a nuisance, and radio announcers joked that listeners should make their morning toast by rapidly rubbing two pieces of bread together. But it soon became no laughing matter as silent computers, dark traffic lights, and cold stoves gave rise to an outraged citizenry. This is not to mention the far more serious damage to the South African economy from the shutdown of mines crippled by the failures, shopping malls turned into ghost towns, and other assaults that experts estimated would cap growth at 4 ½ percent, far less than the level the government deemed necessary to reduce the country's 25 percent unemployment rate.

The crisis stemmed from a very unhappy combination of government/industry lack of communication and follow-through on a 1998 White Paper, which argued that at the rate the economy was growing South Africa faced a serious power shortage by 2007 unless actions were taken to expand energy-supply capacity.

The Mbeki government took office the following year and was unsuccessful in its attempt to get private investors to finance construction of new power plants. Only later did the government give Eskom, the state power monopoly, permission to begin expanding capacity. But by then it was too late, much too late, since power plants don't magically appear overnight. Generally speaking, it takes at least five years or more to put up a plant and get it online. As the respected South African analyst William Mervin Gumede put it, "The warnings were well known, but the government was too aloof and arrogant to act. This is simply disastrous for the economy."

Complexity mismatch shows a different face in this situation, as the predecessor to the Mbeki government recognized that the economic system was growing (becoming more complex) at a rate far exceeding the complexity of the country's power grid. The only outcome of such an increasing gap would inevitably be a massive power shortage. And, of course, that's what it took to close the gap in the form of the Mbeki administration finally caving in and allowing the state-controlled electrical utility to take action to beef up its power-generation capacity. But it took an X-event in the form of the wave of power outages in 2008 to bring that complexity gap back to a level at which the country could function again.

Meanwhile, South Africans fumed and fretted at the daily disruptions to their lives, as elevators stalled between floors, shops closed down, petrol stations were unable to pump gasoline, traffic lights were disabled, and restaurants left food half cooked in their ovens. What's the solution? Well, as one engineering consultant put it, "Because of this situation, economic growth just stops. In that way, the problem solves itself."

While the appalling situation in South Africa was unusual in the combination of its systemic nature and basic human failings, the loss of electrical power for varying periods of time is commonplace all over the world. Here is a telegraphic account of a few more incidents of this type over the past few years

- In February 2008, a fifth of Florida's population lost power at various times as a relatively minor glitch in the power grid caused a nuclear plant to shut down. Later investigation showed the cause to be "human error" yet once again, as an engineer looking into a malfunctioning switch at a substation in west Miami disabled two levels of protection for the system. As he was making a measurement of the switch, a circuit shorted out. Normally, the protection system would have contained the problem. But because both levels of protection had been disabled, the short circuit cascaded throughout the system. Officials later said it would have been appropriate for the engineer to disable one of the levels—but not both—and they didn't know why he took it upon himself to remove both levels at the same time.

- On Saturday night, November 5, 2006, about ten million people in France, Italy, and Germany were trapped in trains and elevators when electric power went out for about half an hour. The German energy company E.ON said the problem began in northwestern Germany, where its network became overloaded due to the shutdown of a high-voltage transmission line over a river to let a ship pass safely. The company said it had carried out similar shutdowns in the past without incident, and that it could not understand where and why the shortfall in electricity began. The German government immediately demanded an explanation from E.ON of what happened—and how it would prevent a reoccurrence. According to the German economy minister Michael Glos, "Power outages of this kind are not only annoying for people, but also represent a considerable risk for the economy."

- In January 2008, a cat in Nampa, Idaho, picked the wrong place to come in out of the cold, causing twelve thousand homes and businesses to lose power. It seems the cat entered a power company substation, curled up next to a warm transformer, and contacted a live circuit, creating a short that blew out lines. The short circuit also short-circuited the last of the cat's nine lives. (Reports do not say whether the cat was black or not.) The power outage also disabled traffic lights in the city before power was restored a few hours later.

- In September 2007, the US Homeland Security Department showed a video graphically displaying destruction caused by hackers gaining control of a crucial part of the electrical grid. Specifically, the hackers modified the operating code for an industrial turbine, causing it to spin wildly out of control until it broke apart, flinging bits and pieces of smoldering metal inside the power-generating unit, ultimately creating a huge fire. Fortunately, this attack never happened as the video was a "What If?" type of exercise showing the potential for disruption of the power grid by terrorists bent on bringing down one of the many almost totally unguarded components of the power supply network.

- September 8, 2011, saw another major blackout in the United States, when more than five million people throughout San Diego County, Baja California, and parts of Arizona lost electricity beginning around three o'clock in the afternoon. A high-voltage transmission line from Arizona to California failed, setting off a cascade of smaller failures that ultimately knocked out the San Onofre nuclear power plant. When that went down, so did electrical power throughout the region. Power was out only for a day or so from this incident, not long by the standards of many electrical failures. But it could not have happened at a worse time of the year, as the outage coincided with the final day of a heat wave that saw temperatures rising into the upper 90s and low 100s (Fahrenheit) inland, and as high as 114°F in the Southern California deserts. As local resident Kim Conway stated, "It's the worst day of the year—it's just freaky that it was so close to 9/11." Her cart, incidentally, was filled with beer and alcohol as she prepared for an impromptu block party with her neighbors.

And so it goes. The point is that power grid failures happen all the time—everywhere. And they will continue to happen all the time everywhere, for reasons great and small. What's important is to understand how the grid can be made more reliable and resilient to the type of failures that shut off power to tens of millions of people for days at a time. No system is going to be proof against human stupidity or wandering cats. But we can and must do a lot better than in the past. With the foregoing examples in mind, let's look at just what role the electrical power grid plays in everyday human affairs.

To get a feel for what's at stake when the lights go out, we can do no better than to look at the specifics of how the loss of power impacted people in the San Diego area after the X-event I described a moment ago:

- No reports of crime occurred, but several traffic accidents happened due to traffic signals being out of service, which in turn led to massive gridlock on major thoroughfares. Gas stations closed, as did other retail outlets like supermarkets.

- Some residents had to pry open their garage doors to get at their cars, as the electrical motor operating the door would not function.
- All forms of public transportation were on a hit-or-miss basis, depending on whether electrical power was needed to operate the service (trains and planes) or not (buses).
- Hospitals and other emergency facilities continued to operate— but only on backup generators having a limited lifetime before stopping, either from battery failure or lack of fuel.
- Water shortages were reported due to several pumping stations being out of service.

Many would say that this list is hardly catastrophic. And they would be right. The real catastrophe sets in when the services listed here are out for days, maybe even weeks, with no one being able to say when the power will be restored—or even *if* it will be restored. That's when the rioting, looting, and other primitive survivalist-style behavior sets in. The San Diego story is simply a dress rehearsal for an ugly, probably deadly, event due to take place just about anywhere electrical power is used to underwrite people's everyday style of life. And that means just about anywhere in the industrialized world.

THE MATRIX OF LIFE

IN THE FILM *THE MATRIX* EVERYONE IS PLUGGED INTO AN OMNIPER-vasive computer program that orchestrates all aspects of their lives. But the "machine" underlying this virtual-reality reality develops a glitch, threatening the existence of the entire population. It's of more than passing interest to ponder this cinematic world and its crisis in connection with the everyday world we all occupy. Now, though, the "machine" is "the grid"—the electrical power grid—as our daily lives are every bit as determined by the vagaries of this machine as the lives of Neo and his friends are fixed in the matrix.

Jason Makansi, executive director of the Energy Storage Council, has graphically illustrated the foundational role played by the electrical power system in his illuminating book *Lights Out*. When the electricity stops, traffic lights go out; cell phones cease to work; elevators stall between floors; pumps stop pumping water, gasoline, and other fluids to where they're needed; computers shut down; and trains stop running. In short, everyday life is set back to a preindustrial level. So how did we get into such a precarious state, anyway? After all, a hundred years or so ago streets were still lit by gaslights, and local transport was by horse-drawn carriage. So the electrification of society is a relatively new phenomenon, something that came upon us just during the last century. Much like its gold-dust twin, cheap energy from oil, electricity owes its place in our lives to the genius of two well . . . geniuses, both plying their trade as geniuses a little over a century ago in New York City.

The first brilliant innovator is a household name everywhere: Thomas Edison, inventor of the phonograph, lightbulb, and many other commonplace items still used today. In the latter part of the nineteenth century, Edison developed a direct current (DC) linkup of lightbulbs powered by a generating station on Pearl Street in lower Manhattan. Interestingly, even though Edison's system of electricity distribution lost out in the long run to a competing "brand," to this very day there are still about two thousand customers in Manhattan who receive DC power from the successor to that long-ago station.

The problem with Edison's DC system lay in transmitting the power from its source to its consumer. The voltage required to transmit DC power any appreciable distance is just too great for the system to be useful. The voltage of alternating current (AC) power, on the other hand, can be raised or lowered by transformers, making its transmission over vast distances relatively straightforward. Enter our second genius, the somewhat mystical Croatian inventor Nikola Tesla, probably the only inventor who could be put into the same class as Edison in terms of the number and significance of his inventions.

Tesla, who spoke of his insight into the mechanical principles of the motor as a kind of religious vision, once worked for Edison. But when he asked Edison for permission to do research on AC, specifically an AC motor, Edison rejected the notion. Tesla then resigned himself to research on DC. He told Edison that he thought he could improve the DC generator dramatically, whereupon Edison told him that he'd get a $50,000 bonus if he succeeded in this task. After a lot of hard work, Tesla produced a set of twenty-four designs for components that would improve the DC generator substantially, just as he'd claimed. But when he asked about the bonus, Edison told him that he'd been joking. "You don't understand American humor" was the great man's cheapskate excuse.

Hugely disappointed, Tesla quit working for Edison and started his own firm to develop his AC ideas. In 1888, he patented an AC motor, which opened the door to cheap and efficient long-distance transmission of electricity. The industrialist George Westinghouse immediately bought up Tesla's patents. After a few years of false starts, feuds with Edison, and other hiccups, the Westinghouse-Tesla AC system won out over Edison's DC system, paving the way for AC power to become the basis for what is now the entire North American power grid. The key point is that the transmissibility of AC power over great distances is what enabled the grid to become centralized, thus vulnerable to the types of disruptions seen in the examples above. I'll return to this matter in a moment. For now, let's have a deeper look at a number of vulnerabilities that the choice of an AC system has left us open to.

POWER AT RISK

LONG, LONG AGO IN A COUNTRY FAR, FAR AWAY (1970s USA), THERE was a power grid that provided ultrareliable electrical power service at a reasonable, if not bargain basement, price. But the profits were not enough for greedy operators and even greedier politicians, so the free marketeers engineered the fantasy that privatization and deregulation of the power-supply network was just what the doctored ordered to bring even cheaper, more reliable electricity to the masses. But a funny thing happened on the way to deregulation: the power transmission grid got lost! Makansi has identified the following vulnerabilities in the power grid that brought us to its current sad state.

Vulnerability 1: A Deteriorating Transmission Grid: Investment in the transmission system was shunted aside in favor of improving other "sexier" and more visible parts of the system. So the basic infrastructure of the system was actually allowed to deteriorate. The consequence of this lack of investment in transmission is that many of the supposed benefits of deregulation could not take place since they relied on a transmission network that could reliably transport vast amounts of electricity from where it is cheap to where it is needed. As former secretary of energy Bill Richardson described it, the United States was "A major superpower with a third-world electrical grid."

Vulnerability 2: A Much Lengthened Supply Line for Fuel Sources: The energy source for most new power plants in the past decade has been liquefied natural gas (LNG). The next round of plants will almost surely be nuclear powered. The sources for both the LNG and nuclear fuel lie far beyond the shores of the United States. Much of the LNG comes from places like Iran, Russia, or Africa, hardly reliable partners in the current geopolitical scheme of things. Nuclear fuel is in somewhat better shape, being mostly supplied by rather more friendly nations like Canada and Australia. But these are still places many thousands of miles away from where the fuel is actually needed. So if the current trend of outsourcing energy sources continues, a significant fraction of the American electricity supply will be produced by energy sources a great distance away.

Vulnerability 3: Electricity Cannot Be Stored: Unlike oil, electricity cannot be stored in the ground as insurance against supply disruptions or other sorts of rainy days. Of course, it can be stored as chemical energy in a battery, mechanical energy in a flywheel, and the like. But not as electricity. That can be done only with a capacitor, which is fine for small quantities of electricity but nothing like what's needed to power a town.

Vulnerability 4: A Lack of Specialized Workers to Maintain and Operate the Power Infrastructure: In today's MTV and Facebook world, young people aspire to sexy occupations like media consultants, financial manipulators, psychotherapists, legal beagles, and the like. They certainly do not aspire to be engineers, and even those who do look to things like nanotechnology, computers, and other so-called growth industries, not electrical power. As evidence for this deplorable trend, a recent OECD study shows that the fraction of students in the United States taking engineering or science majors is just 15 percent, as compared with 37 percent in Korea and 29 percent in Finland. It's a frightening statistic to hear that currently for every two workers set to retire from the power industry, there is less than one person set to replace them. Moreover, the above figures suggest that the pipeline of new talent from the universities seems ready to supply only a trickle when what's going to be needed is a flood.

Vulnerability 5: The Connective Structure of the Power Grid: The North American power grid is what network theorists term a "scale-free" network. This means it has a few major hubs with lots of minor spokes. So if a random event takes out a component, the chances are overwhelmingly high that the outage will not percolate through the entire system. But if that event knocks out one of the major substations, as it did in the Great Northeast Blackout of 2003, the entire system in half the country is at risk. Given this structure and the critical role played by electricity in the lifeblood of the country, much more attention needs to be directed to the protection of the "hubs."

Vulnerability 6: The Large Environmental Impact of the Grid: Both coal and natural gas are known generators of greenhouse gases, CO_2 for coal, methane for natural gas. That's the good news! The bad news is that methane is twenty times worse than carbon dioxide as a warming agent, and as we saw above, almost all new power plants are LNG fired, not coal. As the LNG pipelines to the plants sometimes extend for thousands of miles, leaks occur releasing methane molecules into the atmosphere. Estimates are that from 2 to 10 percent of the methane escapes as the LNG moves from the storage tanks at the port to the power plants where it's used.

All these vulnerabilities are serious. And most (but not all) can easily lead to a massive failure of the electrical power grid if left unchecked. Currently, there is a lot of talk but little action being taken by either government or the power industry itself. It's no exaggeration to say the system is in crisis and you can believe it's the case when PJM, one of the largest power operators in America, calls the need for a new transmission grid an "emergency." In short, Richardson's "third-world grid" is in critical danger of degenerating into an "out of this world" grid.

If any infrastructure rivals electrical power in its impact on our daily lives, it is surely the system that delivers water to our homes and companies. Whether it's water for drinking, cooking, cleaning, sewage, irrigation, cooling, or myriad other uses, fresh, clean water is truly the sine qua non of modern life. We can live without electricity; we *have* lived without electricity for many thousands of years. But we can't live more than a few days without water. With this in mind, let's see how close we are to an X-event that will turn off the faucet.

II. DRY

EVEN WORSE, A LOT WORSE

EARLY 2004 SAW AN UNHERALDED DEAL BETWEEN TURKEY AND Israel that strikes to the heart of what could turn into a problem facing humankind that may be far more immediate, severe, and life-threatening than either global warming or massive power failures. The agreement between the two countries calls for Israel to send weapons to Turkey in exchange for fresh water to be delivered by tanker to Israeli ports on the eastern Mediterranean. Almost unknown to the world, Turkey is awash in fresh water; it already delivers it by tanker loads to Cyprus and has plans to sell it to Malta, Crete, and Jordan as well.

There is a rapidly increasing need for fresh water worldwide. Every person in the world needs it. And those who don't get it die. It's just that simple. Black or white, life or death. No shades of gray in this story. In contrast to global warming, which some skeptics claim may or may not be happening, the water-supply problem definitely *is* happening. And it's happening *now*. As living standards rise in developing countries and populations increase, the need for fresh water moves in lockstep with these trends. For instance, the OECD estimates that each American uses about six hundred liters of fresh water per day for drinking, bathing, washing, and other household activity. At the other end of the scale, in Mozambique the figure is about twelve liters per day—fifty times less. If the water is not available, things can get very ugly very fast, as events in the United Kingdom showed in 2007.

July is often a rainy month in the United Kingdom. And never more so than in the summer of 2007, when the worst flooding in sixty years hit the West Country and the Thames Valley, as an estimated 90,000 gallons (over 325,000 liters) of water a second was pouring down the Thames the night of July 23 on its way to Oxford, Reading, and Windsor. In the area most severely affected by the floods from the Avon and Severn rivers, homes were left without running water, leading to panic buying of bottled water and food in supermarkets. One twenty-six-year-old woman from Gloucester, a mother of two children, said she had driven more than fifteen miles to buy water after the closure of a local water treatment plant due to the flood. "We have been to three supermarkets and water had sold out in all of them," she reported. "The queues outside the supermarket are horrendous. Everyone is desperate to get their hands on some water. We have heard stories of grown men pushing kids out of the way to get to bottles of the stuff. It is disgusting."

An overdose of water after a few days of flooding in the United Kingdom or elsewhere in the world is indeed a disaster, sometimes even a catastrophe as in New Orleans in 2005. But what about no water at all for a few weeks or a few months as happened in the drought in 2011 in Texas, which saw the two main lake reservoirs serving Austin and other towns in the area dropping to less than 40 percent of their storage levels, a condition that is already rated "Severe" and just a stone's throw from "Emergency"? Similar droughts brought agriculture to its knees in Russia in both 2010 and 2011.

If these situations can result in such huge disruptions in people's lives from what amounts to a temporary shortage of water in a relatively small area of a major industrialized country, what can we expect if there is a prolonged shortage over a large area? Even a partial answer would be nothing pretty, that's for sure. The drought in Somalia, which up to now has killed at least thirty thousand children and affected over twelve million people, is a case in point. As noted earlier, everyone needs water. And if you don't get it, you die. So how do we stand globally? How limited is the supply of fresh water? What is the trend over the coming years and decades?

WHEN THE PIPES RUN DRY

THE UNITED NATIONS ENVIRONMENTAL PROGRAMME PROJECTS water scarcity worldwide in 2025 on a country-by-country basis. According to their study, 1.8 billion people will live in the "red zones" of the world, those areas having a physical water scarcity. This means they will not have enough water to carry on their current level of per capita food consumption using irrigated agriculture, and at the same time meet water needs for domestic, environmental, and industrial activities. To deal with such needs, water will have to be diverted out of agriculture, which in turn leads to the need to import food in these areas.

Equally frightening are the statistics showing the change in water withdrawal as a percentage of available water over the thirty-year period 1995 to 2025. In 1995, only a handful of countries in the Middle East, North Africa, and around the Caspian Sea were taking out 20 percent or more of the water available. But by 2025 not only will all these regions be in the 20 percent and up category, so will all of China, most of Western Europe, *and* the USA and Mexico. By that time, only South America, Russia, West Africa, Canada, Australia, and New Zealand will be in the "safe zone," those regions withdrawing 10 percent or less of the available water.

These numbers show better than any words how acute the water problem is on a global scale. As the final nail in this particular coffin, the number of people affected by this worldwide water shortage is projected to rise from about five hundred million in 1995 to nearly seven billion by 2050.

So the story of fresh water availability in the next few decades if current assumptions about water usage, population, and economic growth are valid is a very grim tale indeed. Provided these assumptions hold up (and don't become even worse), we're in very deep trouble. Can we get out of it? Probably not. But certainly not if we don't get a better understanding at the individual level of just how much water it takes to carry on our current lifestyle. One good way of forcefully bringing home this consumption issue is to examine the so-called virtual water contained in just about every food item we eat, and see how that water usage is translated into the overall water "footprint" of a country.

The concept of virtual water was introduced in 1993 by British water researcher Tony Allan to measure how water is embedded in the production and trade of food and consumer products. Allan argued that people consume water not just in what they drink directly or in what they use for showers. If this were the only usage of water by individuals, the world would certainly not have a water problem. But Allan showed, for instance, that about 140 liters of water stands behind the cup of coffee that we drink in the morning, 2,400 liters of water goes down the hatch with the Big Mac we have for lunch, while a whopping 22,000 liters hides behind the production of the one kilo slab of roast beef put on the table for Sunday dinner. To put these figures into perspective, the 140 liters of water in that morning cup of coffee is about the same amount as the water used by the average person in England per day for drinking and other daily needs.

One might think of a kind of global "trade" in virtual water, whereby countries with little water (think Saudi Arabia) import high-water-consuming products while exporting low-water-consuming products (oil), making water available for other purposes. It would seem a useful act of public awareness to label products with their virtual water content in order to increase people's water awareness. For instance, one ton of wheat contains twelve hundred cubic meters of virtual water, while a ton of rice has twenty-seven hundred cubic meters, nearly triple that amount. Thus, you're saving a lot of water by consuming bread instead of rice. And when it comes to meat products, beef is the real water killer, having nearly three times as much virtual water content per volume as compared with pork, and over five times more than chicken. So the notion of virtual water import via food is an alternative water "source" that acts to reduce pressure on water resources in arid regions.

But public awareness of water content in the foods we eat is certainly not going to cure the growing global water shortage problem by itself. Much, much more needs to be done—soon. Otherwise, twenty years from now the only thing coming out of the faucet when we turn it on will be a lot of hot air, a commodity that's unfortunately never in short supply, especially in the political arena.

ADDING IT ALL UP

ELECTRICITY AND WATER ARE BOTH FLUIDS, METAPHORICALLY IN THE first case, literally in the second. And they are both critical for sustaining life as we know it. To do that, both have to flow from where they're plentiful to where they're not. A further similarity is that the catastrophes associated with both are primarily local, not global. So why are they treated in this book? To answer that, I have to explain what I mean by "local."

A local failure of, say, the electrical power grid means a failure restricted to a particular geographic region. So, for example, the 1977 power failure in New York was very local, affecting only New York City and a few areas immediately north of the city. By way of contrast, the 1965 blackout impacted a geographic region encompassing much of the northeastern United States, a vastly larger geographic area but still "local" when measured against a global power failure or even one that would black out the entire United States. Thus, by the standards of a *global* pandemic, power failures by their very nature are localized to a given geographic area and will never be truly global.

But there is also temporal locality, an event localized in time. This is what we see with the problem of water shortage. In terms of geographic space, the water problem is definitely global; it will affect everyone, everywhere. But not all people will be affected at the same time. As we saw above, even today the water shortage problem is causing havoc for hundreds of millions of people. So in the temporal sense you might say the catastrophe has already occurred. It's just that most of us in the developed world simply don't see it because we are not affected—yet.

A failure of either the power grid or the supply of fresh water would be catastrophic, having a huge impact on the way of life of literally hundreds of millions, if not billions of people. *That* is the reason I've included them in this book.

TECHNOLOGY RUN AMOK

INTELLIGENT ROBOTS
OVERTHROW HUMANITY

IT'S THE LAW

IN THE APRIL 19, 1965, ISSUE OF *ELECTRONICS MAGAZINE*, ENGINEER Gordon Moore, later cofounder of Intel Corporation, wrote the following prophetic words about advances to be expected in semiconductor technology:

> *The complexity for minimum component costs has increased at a rate of roughly a factor of two per year. . . . Certainly over the short term this rate can be expected to continue, if not to increase. . . . [B]y 1975, the number of components per integrated circuit for minimum cost will be 65,000. I believe that such a large circuit can be built on a single wafer.*

A few years later, semiconductor pioneer and Caltech professor Carver Mead dubbed this statement "Moore's law," a term that techno-futurists and the media have now enshrined as the definitive statement underpinning technological advancement in the age of machines. Subsequent mutations, modifications, and tinkering have led to the folk wisdom that what Moore said is that transistor packing/ computer memory capacity/computer performance per unit cost/ . . . will "double every 18 months." Moore actually said no such thing. But what his statement quoted above does state is the general claim that *something* absolutely central to digital technology improvement will increase at an exponential rate—with no increase in cost. Moreover, this rate will continue for at least several decades, if not longer.

Despite the rather grandiose labeling of Moore's observation as a "law," there is actually nothing at all remarkable about what he claimed. In fact, it is a statement that applies with equal force to the overall life cycle of just about any new technology. When a technology is in its infancy, struggling to shove the current competition off center stage, its market share is very small. As the newcomer gains adherents and begins to make serious inroads into the market, the rate of growth increases exponentially. It then peaks and starts the downhill slide leading to this new technology itself being replaced by the "next big thing."

Many studies have shown that the life cycle represented by rate of growth (say in number of units of the product sold per month) obeys very closely the well-known bell-shaped curve of probabilities discussed in Part I. And if we measure the *cumulative* growth of the technology—that is, the fraction of its ultimate total market share achieved over the course of time—that cumulative market share follows the familiar S-shaped curve governing many living and lifelike processes. The high-growth part of the S-curve displays exactly the exponentially increasing growth pattern claimed by Moore for semiconductors.

Even though Moore's law is neither a law nor an extraordinary insight into the growth of a new technology, it is extremely important in the historical development of digital innovations, serving as a kind of goal for an entire industry. The reason is that research and marketing arms of major players in the industry actually believed the forecasts coming from "the law," beliefs that drove them to furiously develop new products aimed at attaining the predicted performance capabilities since they were convinced that their competitors would soon produce the product if they did not. So in a certain sense one can think of Moore's law as a self-fulfilling prophecy. An obvious question is, What are the limits of this principle?

A good place to start in addressing this question is with Gordon Moore himself, who stated in a 2005 interview that the law cannot be extended indefinitely. At that time, he said, "It can't continue forever. The nature of exponentials is that you push them out and eventually disaster happens." In the same interview, Moore also noted that "Moore's law is a violation of Murphy's law. Everything gets better and better." But "forever" and "a long time" are very different things, and so researchers, such as MIT quantum computing expert Seth Lloyd, see the limit as being on the order of *six hundred years*!

In this spirit, speculative futurists like inventor Ray Kurzweil and mathematician, computer scientist, and science-fiction author Vernor Vinge have conjectured that continuation of Moore's law for only another few *decades* will bring on a so-called technological singularity. In his book *The Singularity Is Near* (2005), Kurzweil suggests there are six epochs to the process of evolution, starting with the emergence of information in atomic structures and moving to the state we're in today in Epoch 4, where technical products are able to embody information processes in hardware and software designs. Kurzweil believes that we are currently at the forefront of Epoch 5, which involves the merger of machine and human intelligence. Put another way, this is the point where computer hardware and software manages to incorporate the methods of biology—primarily self-repair and replication. These methods are then integrated into the human technology base. The "singularity"—Epoch 6—occurs when the knowledge embedded in our brains is merged together with the information-processing capability of our machines.

In a 1993 article, Vinge called the Singularity "a point where our old models must be discarded and a new reality rules." Since we humans have the ability to internalize the world and ask "What if?" in our heads, we can solve problems thousands of times faster than evolution can do it with its shotgun approach of trying everything and sorting out what works from what doesn't. By being able to create our simulations at a vastly greater speed than ever before, we will enter a regime so different that it will be tantamount to throwing away all the old rules overnight.

It's of more than passing interest to see that Vinge credits the great visionary John von Neumann for seeing this possibility in the 1950s. Von Neumann's close friend, mathematician Stan Ulam, recalls in his autobiography a conversation the two of them had centering on the ever-accelerating progress of technology and changes in human life. Von Neumann argued that the rate of technological progress gives rise to an approaching "singularity" in human history beyond which human affairs as we know them cannot continue. Even though he didn't seem to be using the term "singularity" in quite the same way as Vinge, who refers to a kind of superhuman intelligence, the essential content of the statement is the same as what today's futurists have in mind: an ultraintelligent machine beyond any hope of human control.

Radical futurists claim that this merger between the human mind and machines will enable humankind to surmount many problems—disease, finite material resources, poverty, hunger. But they warn that this capability will also open up unparalleled possibilities for humans to act on their destructive impulses. For those readers old enough to remember the golden age of science-fiction films, this is all eerily reminiscent of the marvelous 1956 classic *Forbidden Planet,* in which intrepid intergalactic explorers discover the remains of the Krell, an ancient civilization that possessed the power of creation by pure thought alone. The Krell apparently vanished overnight when the destructive power of their alien Ids was given free rein. If you happened to have missed the film, a reading of Shakespeare's *Othello* makes the same point.

It's important to note at this juncture that Kurzweil's argument does not depend on Moore's law remaining in effect indefinitely, at least not in its original form pertaining just to semiconductors. Rather, he believes that some new kind of technology will replace the use of integrated circuits, and the exponential growth embodied in Moore's law will then start anew with this new technology. To distinguish this generalized version of Moore's law, Kurzweil has coined the term "the law of accelerating returns."

The type of X-event I focus on in this chapter involves the emergence of an "unfriendly" technological species whose interests conflict with those of us lowly humans. In such a planetary battle, the humans might win out. But it's not the way to bet. So let's look into the arguments for and against this type of conflict, and see if we can get some insight as to why the radical futurists think we should be wondering and worrying about these matters at all.

THE GNR PROBLEM

THERE ARE THREE RAPIDLY DEVELOPING TECHNOLOGIES THAT CONcern most "singularity theorists" like Kurzweil, Vinge, and others. They are genetic engineering, nanotechnology, and robotics, which taken together form what is often termed "the GNR problem." Here's a bird's-eye view of each.

- *Genetic Engineering:* In the past decade or so, the manipulation of the DNA of plants and animals opens up the possibility of producing organisms having specific properties deemed desirable by the "breeders." These might be practical things like disease-resistant tomatoes or bigger, fatter chickens. Or they might extend to breeding more attractive or smarter humans. In any case, people are concerned about this kind of advanced genetic modification getting out of control and leading to a runaway flood of species that could push humankind off the planet.

- **Nanotechnology:** Huge research efforts are under way to provide practical means for controlling matter at the molecular, or even atomic, level. The catchall term "nanotech" is used to describe a cluster of such efforts, which include things like the use of engineered molecules to clean out clogged arteries (nanomedicine), the employment of molecules as switches in electronic devices (nanoelectronics), and the construction of atom-sized machines to assemble totally novel sorts of products (nanomanufacturing). Ethicists and futurists worry about the possibility of these nano-objects attaining the capability to manufacture copies of themselves, leading to a cascade of "nanobots" flooding the planet.

- **Robotics:** The past decade or two have seen the development of machines performing specific functions like welding parts for automobiles or vacuuming the floor in your house. What is not at all common is a machine driven by a computer program outside the control of the programmer possessing the ability to think like a human being. Moore's law suggests that computer hardware is approaching the point where such an artificial intelligence can be realized.

All three of these threats give the same apocalyptic vision: a technology run amok that's developed beyond human control. Whether it's genetically engineered organisms pushing nature's creations off center stage, a plague of nano-objects vacuuming up matter to leave waste products—a kind of "gray goo"—coating the entire planet, or a race of robots breeding like hyperactive rabbits to force humans out of the evolutionary competition, the common factor underwriting each of these dark visions is the heretofore unseen ability of engineered technology to replicate. Killer plants breeding copies of themselves, nano-objects soaking up whatever resources they need to make more and more nano-objects or robots building more robots, all lead to the same unhappy end for humans: a planet that can no longer sustain human life, or what's worse, a planet in which we humans can no longer control our destiny but have been usurped by objects generated by our own technology.

Up to now, a potentially dangerous technology like a nuclear bomb can be used just once—build it and use it. Then we humans have to build it again. Technologists argue that genetically engineered organisms, nano-objects, and robots will be free of this constraint. They will be capable of self-reproduction on a speed and scale never before seen on this planet. When that crossover point is reached, the curtain starts to fall for humankind as the dominant species on the planet. Or so goes the scenario painted by techno-pessimists like Bill Joy, cofounder of Sun Microsystems, who argued in 2002 that we should impose severe restrictions on research in these areas in order to short-circuit this kind of technological "singularity." I'll take up those arguments, pro and con, a bit later.

Now let's look at one of the more interesting threats of the foregoing type, a plague of robots, as a viable candidate for relegating us humans to the scrap heap of history.

INTELLIGENT MACHINES

ALMOST FROM THE VERY INCEPTION OF THE MODERN COMPUTER ERA in the late 1940s, the idea of the computer as a "giant brain" has been a dominant metaphor. In fact, early popular accounts of computers and what people claimed they would be able to do refer to them as "electronic brains." This metaphor gained currency following a now-legendary meeting at Dartmouth College in 1950 on the theme of what we now call "artificial intelligence," the study of how to make a computer think just like you and me. At about the same time, British computer pioneer Alan Turing published an article titled "Computing Machinery and Intelligence," in which he outlined the case for believing that it would be possible to develop a computer that could think, human style. In this article, Turing even suggested a test, now called the *Turing test,* for determining if a computer was indeed think-

ing like you and me. The Turing test says the computer is thinking
like a human if an interrogator cannot reliably decide whether the
machine is a human or a machine through a sequence of blind inter-
rogations, in which the interrogator cannot see the object being questioned. What is relevant here is that for a race of robots to take over
the world, they must have some way of processing information about
the physical world received from their sensory apparatus. In short,
they need a brain.

The issue is whether technology has come to the point at which a
brain sufficient for the job can be put together from the kind of computing equipment currently on offer or to be on offer soon. (Note: It is
not required that the robot be able to solve all the same problems that
humans encounter. Nor is it necessary that the robot think *in the same
way* as a human. All that's needed is that a brain be good enough to
give the robot a survival advantage in competition with humans.) But
for the sake of comparison, let's confine our attention to the question
of how much computing power we need in order to match or exceed
the computational capacity of the human brain.

First we consider the brain. From numerous studies focused on
estimating the processing required to simulate specific brain functions like visual perception, auditory functions, and the like, we can
extrapolate the computing requirements for the particular part of
the brain involved to the entire brain by just scaling up. So, for
instance, estimates suggest that visual computation in the retina requires roughly 1,000 million instructions per second (MIPS or cps).
As the human brain is about 75,000 times heavier than the neurons in this part of the retina (about one-fifth of the entire retina,
weighing approximately 0.2 grams), we arrive at an estimate of 10^{14}
instructions per second for the entire brain. Another estimate of the
same sort can be obtained from examination of the auditory system.
It leads to a figure of 10^{15} cps for the entire brain. All other such exercises have arrived at more or less this same number as a reasonable
estimate of the power of a single human brain.

How does this compare to a computer? Today's personal computer provides about 10^9 cps. By the law of accelerating returns, we can expect this figure to be stepped up to that of the brain in about fifteen years—or less! So much for processing. What about memory?

Estimates indicate that a human who is expert in some domain such as medicine, mathematics, law, or chess playing can remember around ten million "chunks" of information. These chunks consist of pieces of specific knowledge, along with various patterns specific to the domain of expertise. Further, experts say that each such chunk requires about one million bits to store. So the total storage capacity of the brain comes to around 10^{13} bits. Estimating memory requirements in the brain by counting connections between neurons leads to a larger figure of 10^{18} bits of memory for a brain.

According to the technological growth curves for computer memory, we should be able to buy 10^{13} bits of memory for less than one thousand dollars in about ten years. So it's reasonable to expect that all the memory needed to match that in the brain will be readily available not later than around the year 2020.

Putting the two hardware estimates together, we see that we're within twenty years of being able to match the brain's processing and memory capacity with a computing machine costing around one thousand dollars.

Now, what about software? Matching the hardware requirements of the human brain is likely within the next decade or so. But the "killer app" arrives when we can match the computer's speed, accuracy, and unerring memory with human-level intelligence (i.e., software). In order to do this, we have to effectively "reverse engineer" the brain, capturing its software in the hardware of tomorrow.

When it comes to simulating the brain, the first thing we have to note is all the many ways the human brain differs from a computer. Here are just a few of the more important distinctions:

Analog Versus Digital: A modern computer is essentially a digital device that relies on great speed to turn switches on and off at a dazzling rate. The brain, on the other hand, uses a combination of digital and analog processes for its computations. While in the early days of computing people set great store by the seeming digital aspect of the brain's neurons, we have found that human brains are actually mostly analog devices using chemical gradients (neurotransmitters) to open and close its neuronal switches. So to argue a similarity between a computer switching circuit and one in the brain is a big stretch, to say the least.

Speed: The brain is slow; a computer is fast. In fact, the typical cycle time of even a slow computer is billions of times faster than the cycle time of a neuron, which is about twenty milliseconds. Thus, the brain has only a few hundred cycles to do its job of recognizing patterns.

Parallel Versus Serial: The brain has trillions of connections linking its neurons. This high degree of connectivity allows *lots* of computations to be carried out in parallel. This is totally unlike almost all digital computers, which do one operation at a time in a serial fashion.

These are but a few of the features distinguishing a human brain from a computer. But with the computing power that's just around the corner, we will still be in a position to simulate the brain without actually trying to fabricate it. What's needed for the next stage of evolution is for computers to be *functionally equivalent* to the brain, not to duplicate its precise physical structure.

All this having been said, simulating a human brain functionally inside a computer is not quite the same thing as simulating a human *being*. Or is it? Of course, a disembodied suprahuman brain might easily displace carbon-based humans as the dominant "thought processors" on the planet. But even a disembodied brain needs somehow to sustain its existence in some type of material medium. Nowadays that medium is the motherboard, keyboard, monitor, hard drive, RAM memory chips, and all the other hardware of your computer. Tomorrow, who knows? But what we do know is that there will have to be *some* type of physical embodiment of the intelligence. This means a sensory apparatus for accessing the world outside the intelligence, as well as some sort of boundary separating that intelligence from what is "outside." So much for computers. What about robots?

A BRAIN-IN-A-VAT VERSUS ROBBY, THE ROBOT

AS I WRITE THESE WORDS IN MY OFFICE AT HOME, IN THE NEXT ROOM a robot called a "Roomba" is moving about in the living room diligently vacuuming the carpet and floors. My hat's off to the design team at iRobot, Inc., who developed this little gizmo, as it does an excellent job at a task that I hate—exactly what most of us wish for in a robot. Basically, what we usually have in mind is some sort of automaton that will do our bidding with no questions asked, relieving us of various chores and duties that need doing (like vacuuming) but that in truth are pretty tiresome and boring. But what we certainly do *not* have in mind is a collective of intelligent robots who think that perhaps the tables should be turned and that humans should be vacuuming *their* floors them instead of the other way around. What are the possibilities of that flip-flop?

Just before setting the Roomba loose in my living room, a friend and I watched the classic 1956 sci-fi film *Forbidden Planet* that I mentioned earlier. Although the technology envisioned by the film's producers fifty years ago is now a bit antiquated, the story and the moral is as fresh as this morning's croissants from the bakery on the corner.

Although the F/X specialists in the 1950s were not quite up to today's standards, the portrayal of Robby, the Robot, a piece of machinery that serves humans as a driver, cook, transport device, and overall factotum is wondrous. Even my teenage mind was fascinated by the possibilities when I first saw the film, and I marveled at Robby's capability to learn new tasks and comprehend human instructions. Moreover, at the story's denouement Robby remains loyal to his human masters, as all of his wiring short-circuited when he was given instructions to harm a human.

The question raised for us by Robby is whether a robot with those almost superhuman properties can be guaranteed to follow something like Isaac Asimov's laws of robotics. In about 1940, Asimov proposed the following laws that a robot must adhere to in order to remain a servant to humans, not an evolutionary competitor.

> *First Law*: A robot may not injure a human being, or, through inaction, allow a human being to come to harm.
> *Second Law*: A robot must obey orders given to it by human beings, except where such orders would conflict with the First Law.
> *Third Law*: A robot must protect its own existence as long as such protection does not conflict with the First or Second Law.

These laws of robotics stipulate that robots are to be slaves to humans (the Second Law). However, this role can be overridden by the higher-order First Law, which prevents robots from injuring a human, either by their own action or by following instructions given by a human. This prevents them from continuing any activity when doing so would lead to human injury. It also prevents the machines from being used as a tool for or to assist in various types of physical assaults on humans.

The Third Law generates a survival instinct. So in the absence of conflict with a higher-order law, a robot will

- seek to avoid this destruction through natural causes or accident,
- defend itself against attack from other robots, and
- defend itself against attack by humans.

Roger Clarke and others have noted that under the Second Law a robot appears to be required to comply with a human order, so as to (1) not resist being destroyed or dismantled, (2) cause itself to be destroyed, or (3) (within the limits of paradox) dismantle itself. In various stories, Asimov notes that the order to self-destruct does not have to be obeyed if obedience would result in harm to a human. In addition, a robot would generally not be precluded from seeking clarification of the order.

Another gap in Asimov's three laws, and one that is very important for our purposes here, is that the laws refer to individual human beings. Nothing is said about robots taking actions that would harm a group, or in the extreme case, humanity as a whole. This leads to the following:

Zeroth Law: A robot may not injure humanity, or through inaction allow humanity to come to harm.

These "laws" of robotic good citizenship impose severe constraints as to what's needed to keep a band of intelligent robots in check. Given the propensity of intelligent objects to evolve so as to enhance their own survival, it seems unlikely that robots of the sort we're envisioning here will be satisfied to serve humans as they develop the capacity to serve themselves. In the 2004 film *I, Robot,* which is loosely based on Asimov's 1950 collection of short stories, robots reinterpret the laws and logically conclude that the best way to protect humans is to rule them. The paradox here is that to be really useful robots have to be able to make their own decisions. But as soon as they have the capacity to do this, they acquire the ability to violate the laws.

Now back to the vexing question: Will robots take over the world? The short answer is . . . a definite maybe!

One of my favorite rejoinders to the claims of some futurists about robotic takeover in the next few decades is that the bodies of robots will be made of mechanical technology, not electronic. And mechanical engineering technology is simply not developing at the same furious rate as computers. There is no Moore's law in the mechanical realm. To illustrate, if automobiles had developed at the same pace as computers, we would now have cars smaller than a match box, traveling at supersonic speeds, and transporting a trainload of passengers while consuming a teaspoonful of gasoline. In short, size matters when it comes to mechanical technology, and the rule is the bigger it is, the more powerful it is. Computers are just the opposite.

So even if we have robots hundreds of times more intelligent than us in a few decades, humans will still maintain a vast mechanical superiority. Humans will be able to knock over such a robot without breaking a sweat, climb stairs and trees easier than any robot on wheels could hope to do, and generally outperform robots on almost any task requiring the delicate manipulative capabilities that we have in our hands and fingers.

If I were a betting man, I'd put my money on the above argument for human superiority in the mechanical dexterity department. And this despite the fact that we already have robots doing surgical operations by remote control, along with robotic soldiers carrying out missions in regions infested with land mines, poisonous gases, and other hazards to humans. The fact that robots can execute such tasks is indeed impressive. But these are very special-purpose devices, just like the Roomba vacuum, designed to perform a very special job— and only that job.

Humans, on the other hand, have a far greater capacity to deviate from the planned program when circumstances don't quite fit into the predefined framework that the robot's "brain" expects to encounter. Of course, you might argue that when the robot brain begins to surpass the human brain in its information-processing capabilities and in its ability to adapt to unanticipated circumstances, the game may indeed be up for us humans. With this ambiguous prospect in mind for a robotic takeover, let's return to the question of the Singularity and examine *when* it might happen.

THE SINGULARITY

IN THE 1993 PAPER BY VINGE THAT SPARKED OFF THE VOLUMES OF debate about the Singularity, several paths are sketched that could lead to the technological creation of a transhuman intelligence. To paraphrase Vinge, these include:

- Computers are developed that are "awake" and superhumanly intelligent.
- Large computer networks (e.g., the Internet) and their associated users "wake up" as a superhumanly intelligent entity.
- Computer/human interfaces become so intimate that users of the interface are considered superhumanly intelligent.
- Biological science provides the means to improve natural human intelligence.

The first three elements on this list involve improvement in computer technology, while the last is primarily genetic. And all may well rely on nanotechnological developments for their realization. So each facet of the GNR problem discussed earlier makes its appearance in the unfolding of the Singularity. And once such an intelligence is "alive," it's likely that it will lead to an exponential runaway in development of even greater intelligences.

From a human point of view, the consequences of the emergence of these superhuman intelligences are incalculable. All the old rules will be thrown away, perhaps in just a few hours! Developments that previously were thought to take generations or millennia may unfold in a few years—or less.

For the next decade or so we probably won't notice any dramatic movement toward the Singularity. But as hardware develops to a level well beyond natural human abilities, more symptoms of the Singularity will become evident. We will see machines take over high-level jobs such as executive management that were previously thought of as the province of humans. Another symptom will be that ideas spread far quicker than ever before. Of course, we already rely on computers for a bewildering array of tasks, as I described solely in the context of communication in the earlier chapter on the Internet. But even in such a mundane matter as writing this book, I occasionally shudder when I think of what it was like just three decades ago when I wrote my first book—literally by hand! That thought is but a distant early-warning sign of things to come as we approach the Singularity.

And what of the moment when the Singularity actually arrives? According to Vinge, it may seem as if our artifacts simply "wake up." The moment we cross the threshold of the Singularity we will be in the Posthuman era.

The most crucial point here is whether the Singularity is actually possible. If we can convince ourselves that it can indeed happen, then nothing short of the total destruction of human society can stand in its way. Even if all the governments of the world were to try to prevent it, researchers would still find ways to continue making progress to the goal. In short, if something can happen, it will happen—regardless of what governments, or societies as a whole, might think about it. That's the natural way of human curiosity and inventiveness. And no amount of political bombast or hand-wringing morality is going to change that state of affairs.

So assuming the Singularity *can* take place, when is the "crossover" going to occur? There seems to be a reasonably uniform consensus on the answer: within the next twenty to thirty years. The technology futurist Ray Kurzweil has been even more specific. In his book *The Singularity Is Near,* a kind of bible of Singularitarians, he states:

> *I set the date for the Singularity—representing a profound and disruptive transformation in human capability—as 2045. The nonbiological intelligence created in that year will be one billion times more powerful than all human intelligence today.*

That's about as definite as you can get in the forecasting business!

For what it's worth, even though I firmly believe that there will *be* a Singularity, I'm personally rather skeptical about the timing aspect of the whole business. The arguments from Moore's law, accelerating returns, human curiosity, and the like leading to this "grand" event in a couple of decades strikes me as rather reminiscent of the kinds of pronouncements made in the early 1950s by AI advocates about what computers would (or wouldn't) do in the years to come. Some of those claims included becoming the world chess champion within ten years, translating languages at the skill level of first-rate human translators in the same time frame, becoming electromechanical butlers serving dry martinis after a hard day at the office, and so on. Well, some of these goals actually were achieved, such as a computer (Deep Blue II) beating the world chess champion (in 1997, not in the 1960s, and by using methods totally unlike what a human player would employ), while others are as far away as ever from being achieved (high-quality, human-level language translation). In fact, the whole line of argument by the Singularitarians is a familiar one in the futurology business: extrapolate current trends and ignore the possibility of any surprises getting in the way. But, of course, this argument only puts off the day of accounting, and I strongly suspect we will see the kind of superhuman intelligence the Singularity calls for before the end of this century.

ADDING IT ALL UP

THE COMPLEXITY INCREASE IN THE WORLD OF MACHINES IS RAPIDLY outpacing that of the human side of the ledger. In contrast to some of the complexity gaps I've spoken of earlier, such as an EMP attack or an Internet crash, the Singularity is an X-event, whose unfolding time is decades, not minutes or seconds. But its impact will be dramatic and irreversible, pushing humans off center stage in the grand evolutionary drama of life on this planet.

So there it is. Arguments, pro and con, for the end of the Human era. In the final analysis, it seems a good bet that the GNR problem will indeed lead to the kind of transcendent intelligence that the Singularity will usher in.

THE GREAT UNWINDING

GLOBAL DEFLATION
AND THE COLLAPSE OF
WORLD FINANCIAL MARKETS

DOWN AND OUT IN D.C.

IN THE ENTERTAINING FINANCIAL BEST-SELLER *THE DAY AFTER THE Dollar Crashes,* financial analyst and commentator Damon Vickers paints a one-week scenario that leads from business as usual in the currency markets to a full-fledged grassroots campaign for reform of the political system in America. Here are a couple of milestones along Vickers's road from normalcy to lunacy and back again.

Wednesday, 10 A.M.: The US government is holding its regular auction of Treasury bills, asking the world to finance the profligate American way of life. The United States has been borrowing from Peter to pay Paul for decades, so the Treasury anticipates another routine auction of its debt to continue the global-level Ponzi scheme. But to the shock of the government and the financial markets, the world finally stiffens its spine and says, "No way."

Sunday (Evening) to Monday (Morning), New York and Asia: Given the intricate interconnections and high correlations linking the global financial markets (high network complexity), as soon as the US debt fails to find buyers, Asian currency markets fall off the edge, sending markets around the world into free fall. The New York Stock Exchange (NYSE) opens at 9:30 A.M. and closes twenty minutes later, buried in an avalanche of sell orders.

Tuesday, 11:30 A.M.: The NYSE opens two hours late after global markets have sunk by nearly 10 percent from Friday's close.

Friday, 2 P.M.: After a minor recovery lasting from Tuesday through Friday morning, exchanges in Europe and the United States continue to fall even in the face of a big US interest rate increase.

The story gets even uglier, as everyone is selling everything— stocks, bonds, currencies, commodities. Confusion reigns supreme as panic circles the globe. By Wednesday, the IMF and other global financial institutions establish a new world currency exchange system, decreeing that all countries stop printing their national currencies. And so things go. Question: Is this just a fantasy cooked up to titillate the imagination? Or is there a real possibility of something like this type of panic actually happening?

This question is of more than passing interest to everyone who has a bank account, a job, runs a business, or just spends money. The global economy is a wonder to behold, with a global GDP nearing 100 trillion dollars. The financial system of banks, brokerage houses, savings and loan associations, and the like serve as the vehicle by which this huge sea of money makes its way from one place to another as needed. So a freeze-up (or meltdown, take your pick) of the financial system would be like pouring sand into the lubrication system of your car. The car won't go very far without lubrication, and neither will the world economy without the global financial system. Now let's get back to whether the scenario above is a realistic picture of what might happen or just a fantasist's pipe dream.

To answer this, all we really need do is look at the news from the first days of August 2011, changing a few small details of Vickers's scenario to reflect global concerns over the debt crisis in the Eurozone, where Greece was (is?) threatening to default on its sovereign debt, add in the downgrading of US government debt by a rating agency, which in turn spooks investors who now start worrying about the creditworthiness of the United States Treasury, and then sprinkle on a bit of well-founded concern over the willingness of US legislators to act in the nation's best interests instead of their own.

For such a Vickers-like panic to get started, many conditions would have to be satisfied at roughly the same time. These include an ongoing systemic weakness in the US financial system brought on for instance by a debt crisis or a fading economic profile in the United States, an event that serves to set off the panic, and of course, a viable alternative to investment in US stocks and bonds. The default alternative that many traditionally retreat to is gold. But it might be almost any commonly recognized storehouse of value like diamonds, oil, drugs, platinum, or most commonly, hard cash.

Amusingly, even though the investment quality of US Treasury debt in the form of long-term T-bills was downgraded by a major ratings agency, when everything came undone in early August 2011, the preferred safe haven for parking money that investors sucked out of the stock market was, you guessed it, US Treasury bills. This resulted in the price of these instruments shooting up by more than 20 percent in less than two months! But who would have thought otherwise? After all, when markets panic and sellers end up with a fistful of cash in their hands, that money has to go somewhere. Where it ended up was exactly where the conventional wisdom said it shouldn't/couldn't: into US government debt in the form of long-term Treasury bills. When in doubt, deal with the devil you know, in this case the US government.

This fact shows better than any statistic that what counts in the world of finance is trust. In other words, belief. Will the institution that holds your money be there to pay off when you need it? Or will the doors be barred when you and thousands of others come to take home their cash? The US government has been the platinum standard in this regard since the end of the Second World War. Whether that privileged position will remain a few years from now remains to be seen. But for now the US Treasury in Washington, D.C., seems to be the best of a pretty unappealing set of alternatives.

Since the story of a collapsing US financial sector is inextricably intertwined with the fate of the American economy, or for that matter the entire global economy, I focus here on the way complexity overload has led to the ongoing global financial crisis. This story will lead to the account of a deflationary storm brewing on the not-so-distant horizon, an X-event that threatens to send the entire world economy into a tailspin that will take decades to pull out of.

FROM TODAY'S PERSPECTIVE, IT STRETCHES THE IMAGINATION TO think of how much faith the global community put into the hands of the central banks and financial regulators to ward off any threat to the integrity of the world's financial system. As Paul Seabright termed it in an article in *Foreign Policy,* we constructed an "Imaginot Line," a nod to the static defensive fortifications that failed to protect France against German invasion in World War II. Seabright identifies three principal economic defenses against financial crisis, each of which was vulnerable on its own, but which taken as a troika seemed impregnable in 2008.

The first level of defense was *deposit insurance.* This part of "the line" was put in place to protect against the notion that the banking crisis of the 1930s was caused by panic and bank runs by small depositors and retail businesses. Enter deposit insurance à la the Federal

Deposit Insurance Corporation and the problem vanishes—at least for those smaller depositors.

The so-called moral hazard problem, in which banks with insured depositors have no motivation to be careful with how they invest (i.e., loan out) their depositors' money and depositors have no motivation to choose carefully the banks in which they place their funds. This symmetry serves to underpin the second level of defense. The solution to the moral hazard problem was to put in place a complicated structure of financial regulations, which required capital requirements for the banks to prevent them from playing fast and loose with the money of small retail depositors. But those rules did not address large professional investors, who were supposed to be ready to bear their own risks. The safety net was assumed to be failure-proof, which meant that whatever risk was in the system would be borne by others, not by the investors themselves. As late as 2003, Robert Lucas, the Nobel laureate in economics, was telling the American Economic Association that the "central problem of depression prevention has been solved, for all practical purposes."

Finally, the third leg of the defensive triangle, *central banking*. From the 1930s onward the central banks were given the job of keeping prices stable, and as a secondary task to promote economic output and keep a lid on unemployment. In this tripartite "protection racket," the central bank was seen as the court of last appeal that would absorb any cracks in the edifice emerging from either of the first two lines of defense. So what went wrong?

To put it compactly, the fatal flaw in this system was that any problem that came up was seen by each of the three "sheriffs" as falling within the jurisdiction of one of the other two. For example, regulators saw the speculative aspects of mortgage contracts as a problem for the central bank, while the central bank saw it as a problem for the regulators. And no one saw it as a deposit insurance issue. Does this sound familiar? Every problem that came in one of the three doors was immediately shuffled off to one of the other two departments. In short, no one was in charge. This benign, self-protective neglect meant that the very act of reducing apparent risks in fact magnified the true risks dramatically.

The end result was that belief in safeguards that didn't really exist led people to think it was safe to take risks that were in actuality orders of magnitude greater than what they thought to be the case. The shared belief that the authorities had the situation under control was totally misplaced. The resulting metaphoric meltdown of the financial system is quite analogous to the real-world meltdown that took place at Japan's Fukushima Daiichi nuclear reactor in March 2011 we recounted in Part I. This was a case of too little complexity in the control system (the combination of the height of the wall and the generator location) being overwhelmed by too much complexity in the system to be controlled (the magnitude of the quake and the ensuing tsunami).

The financial system collapse arose from the same sort of complexity mismatch. Speculators saw a prolonged period from the 1980s onward in which the markets offered only profits without even the potential for losses arising from taking on too much risk. So financial schemers created a dazzling array of increasingly complicated financial instruments that ultimately even their creators didn't fully understand. Credit default swaps (CDSs) are probably the most well chronicled of these exotic instruments, involving what amounts to insurance contracts that pay off if a particular debt obligation like the bond payment due from a country is not met. Credit default swaps are not true securities in the classic sense of the word in that they're not transparent, aren't traded on any exchange, aren't subject to present securities laws, and aren't regulated. They are, however, at risk—to the tune of $62 *trillion* (the best guess by the International Swaps and Derivative Association). As a result of these "instruments of mass financial destruction," the complexity of the financial services sector shot off to a stratospheric level.

Credit default swaps are not the only contributors to the complexification of the financial services sector. High-speed computerized trading, the deregulation of banks and allowing them to engage in speculative trading, as well as the enormous profits that investment banks and hedge funds have racked up over the last thirty years have each made their own contribution to a level of complexity in the sector that swamps the ability of bankers and traders to fully understand and even try to control.

But what about the regulators, the insurers, and the central banks? By now, you know the answer. The complexity of this control system was actually weakened by legislative action such as the repeal of the Glass-Steagall Act at a time when the complexity of the financial system they were charged to oversee was growing exponentially with each new boutique product served up by the wizards of Wall Street. The central banks and the regulatory agencies had basically the same tools at their disposal in 2007 that they had had for the preceding fifty years. The emerging complexity gap was a disaster waiting to collapse the system. The crisis we're still "enjoying" is the real world's way of rectifying this imbalance, a process that involves painfully wringing out the unsustainable risk and leverage in the financial system.

In case you've been asleep for the past couple of years and didn't notice, the complexity gap between the financial system and its regulators is *still* growing. To see how it will most likely be closed, we need to take a longer and harder look at the way the US economy has been transformed over the past few decades, and how that transformation has given rise to the precarious state we find it in today.

TOO MUCH REALITY

A CHARACTER IN T. S. ELIOT'S PLAY *MURDER IN THE CATHEDRAL* remarks, "Humankind cannot bear very much reality." Although this statement was made in the context of the assassination of Thomas Becket in Canterbury Cathedral in 1170, nothing much about human nature has changed since Eliot penned these words, and for that matter, since the time of Becket's assassination. One such overdose of reality rests at the very heart of the post-2008 Great Recession. A deeper look at the ultimate cause of this financial collapse sheds much-needed light on why a global economic collapse becomes more likely with each passing day.

The conviction in 2011 of billionaire hedge-fund manager Raj Rajaratnam on the grounds of insider trading brought a vast outpouring in the media about the lack of prosecution of those the general public see as the ultimate culprits who perpetrated the crash of 2007–2008. According to media pundits, the public wants the blood of the bloodsuckers on Wall Street. The Rajaratnam trial served to focus this sense of outrage. In the words of economic columnist Robert Samuelson, "The story has been all about crime and punishment when it should have been about boom and bust." In his analysis of the Great Recession, Samuelson goes on to note that the political left and right each have their own set of culprits, but that neither side is really able to make their story stick. Perhaps this is why so few actual "criminals" have been brought to the dock. Rather, the correct reply to the question of who really made the crash is, *We all did*.

The big question is why virtually everyone bought into the boom times and dismissed any naysayer as a Chicken Little? The answer is not hard to come by: very few traders or investors who were active in the market in the years before 2008 had experienced anything other than prosperity. It was a state of affairs that people simply took for granted. This confidence was coupled with an underlying belief that economists—like the Fed's Alan Greenspan or those at the International Monetary Fund and the European Central Bank—had mastered the science of how to maintain a stable economy, and that their expertise would rule out another 1930s-style Great Depression. In short, everyone took a stable, prosperous economy as a taken-for-granted background reality, a state of heavenly economic grace that would persist forever. Samuelson argued, "The most significant legacy of the crisis is a loss of economic control." These thoughts were echoed by the Nobel laureate Paul Krugman, who described the emergency not as one of housing or even economic (mis)management, but a crisis of people's faith in the entire economic system. Investors no longer believe that highly complex, risk-free moneymaking machines like collateralized debt obligations, auction-rate securities, or any of the other fancy financial instruments dreamed up by the "wizards" of Wall Street will function the way they are supposed to. This loss of trust in a system leads to a kind of self-fulfilling prophecy of just the sort described above. *Wow. So it* can *happen here. And it* can *happen again.*

In an article in the *Atlantic* in 2010, Derek Thompson and Daniel Indiviglio, senior editors of the magazine, outlined five ways for the economy to dive in for a double-dip recession—or worse. I've listed them here in the order the authors thought was most to least likely. I leave it to readers to scramble this ordering to suit their own beliefs about the ranking in light of events at whatever time you happen to be reading this chapter.

Housing Falls Off the Cliff: Closely allied to the huge unemployment problem in the United States is the anemic housing market. Weak home sales and increasing foreclosures continue, and perhaps even escalate, putting further downward pressure on housing prices. This in turn makes it very difficult for homeowners to get out from under mortgages they can no longer afford, thus contributing to an even greater number of foreclosures. The end result is that lower home prices encourage people to save more and spend less, leading to a precipitous drop in stocks and a further tightening of credit markets. Ultimately, growth turns negative and the economy teeters on the edge of a massive deflationary spiral.

Consumer Spending Continues to Decline: People's belief in an economic recovery dwindles, and spending slows to a trickle. The stock market sees nothing but gloom and doom, as business revenues flatten out, unemployment continues to rise, and the government does nothing but print more money. The markets start selling off a percentage point or more several days in a row, and as people see their savings disappearing faster than a trickle of water in the desert, they cut back on spending even further. Again, growth turns negative and deflation looms.

The Return of Toxic Assets: In their bailouts, the Treasury Department intended to actually purchase the toxic real estate assets held by the banks. But as they couldn't quite figure out how to do this quickly enough to help, the banks simply took the money—but held the "assets." As home and commercial real estate values continue to fall, so do the values of these toxic assets that still sit in the vaults and on the books of all the major banks. As the assets suffer another round of losses, markets sell off, credit dries up, and growth turns negative yet once again.

Europe Comes Undone: Slow growth in the southern countries of the Eurozone leads investors to demand higher rates of return for the bonds of these countries. This leads to further austerity measures, basically tax increases and spending cuts, which in turn strangles the most important export sources for goods, especially China and even the United States. In a flight out of the euro, the dollar then actually appreciates for a while, further hammering US exports to Europe. Again, the stock market eventually tanks as manufacturing dries up and trade deficits reach an unsustainable level. The American consumer cuts back once again, strangling the domestic market, and (you guessed it) growth turns negative.

Debt, Debt, and Too Much Debt: Uncertainty in the American political process causes buyers of US Treasury notes to demand higher interest rates to balance the risk of an increasingly whimsical Congress. This leads to a reduction of the value of pension and mutual funds holding US government debt, forcing people to save even more and spend less. This dynamic then gives rise to a Hobson's choice of either cutting taxes to promote consumer spending or raising taxes to keep bond buyers happy. Either alternative ultimately leads to a deflationary economic collapse.

In fact, *all* the above gremlins have been jumping out in force over the past year, and the priority ordering for what's going to sink the economy bounces around from day to day like a drop of water on a hot skillet. Right now, the Eurozone debt problem seems to have the upper hand. But who knows what will be flavor of the week tomorrow? Actually, it doesn't matter, because any one of them is quite sufficient to send us into terminal financial and economic collapse.

So much for looking at the recent past and immediate present. We've seen governments in Europe and the United States try to throw money at the problem of the increasing complexity of the financial system without much success in reducing the gap between the system and its regulators. If anything, that gap is widening. So I'll spend a few pages now describing the probable consequences of this failure. What can we expect to see in the near term economic and financial profile of both the United States and the world? This is where things really start to get interesting.

THE INCREDIBLE SHRINKING PIE

RECENTLY, I WAS LISTENING TO A LOT OF SUPPOSEDLY FINANCIALLY savvy talking heads and sifting through a morass of financial blogs, each giving its idiosyncratic postmortem to the unwashed as to why the markets were going down instead of skyrocketing. In this lengthy search for financial nirvana, I decided to look at some of my regular sources, ones that publish what I believe are among the more thoughtful analyses of financial and social happenings. There I found the following statement from Steven Hochberg at Elliott Wave, International, bringing into focus a lot of puzzling aspects about what's taking place right now. Here's what he said in his newsletter on September 8, 2011:

America is downgraded by S&P and one of the greatest American investors in history, Warren Buffett, is put on negative watch by the same ratings agency (the bonds of Berkshire Hathaway). Short-term US government paper is yielding zero. US stocks are crashing and gold continues to rally strongly. According to most US dollar prognosticators, the greenback should be crashing to oblivion. It's not; at least not yet. Instead, the US dollar index remains above . . . the major low established in March 2008, over three years ago. The only explanation for such behavior is deflation.

Did he say deflation? Just about everyone has heard of inflation and many of us know vaguely what it means: increasing prices. But *deflation* is a word that's nearly dropped out of the dictionary over the past few decades. What is it and why is it so important?

Speaking literally, deflation is simply the opposite of inflation: a decline in prices rather than a rise, perhaps together with a contraction of credit and a decrease in the amount of available money. Sounds good—on the surface. Who wouldn't like to see the price of gasoline, iPads, and T-bone steak go down? But like a lot of things that look appealing at first glance, a bit of digging turns up some pretty nasty features we'd like to stay as far away from as possible. Here's why economists and policymakers fear deflation like the plague.

The central problem is what's often termed the "deflationary spiral," a nearly one-way street to no economic growth, no jobs, and very little hope. These are the steps of that precipitous decline that constitutes the deflationary spiral:

1. Prices go down, so companies receive less revenue and make less profit on the sale of their goods and services.
2. Companies lay off workers to adjust for lowered profits, and these newly unemployed workers then spend less money.
3. Companies must now lower prices to tempt cost-conscious consumers back to the cash register, causing prices to fall even further.
4. Go to Step 1 to complete the cycle—but now with even lower prices.

And so it continues, lower prices to fewer consumers to still lower prices to fewer buyers ad infinitum as the entire economy slows to a crawl and ultimately reaches a floor and stops dead in its tracks. This nosedive is extremely difficult to pull out of since those who have money begin adopting the attitude, "Why buy today when prices will be lower tomorrow?"

There are several economic delicacies involving the relationship between labor costs, material costs, time lags, and the like that enter into the details of this story, muddying the waters a bit. But these fine points are unimportant for the issue that concerns us here, namely, the events that cause deflation to get started in the first place. In other words, now that we know what happens when we're in the grips of a deflationary spiral, how does the process actually start?

THERE ARE THREE POSSIBLE (BUT NOT MUTUALLY EXCLUSIVE) PATHS to that fateful first step on the road to deflation:

1. A speculative *bubble bursts,* resulting in a rash of bank failures.
2. Individuals, institutions, and/or national governments *default on their loans.*
3. The central bank *raises interest rates* too much and too fast in an effort to combat inflation.

The end result of any of these paths is that less money is available to be loaned to consumers and to people developing businesses. This means that credit, which is the lifeblood of any modern economy, dries up, so that less money is being spent. That factor, in turn, starts the deflationary spiral. Incidentally, this is the principal reason why governments like the United States turn cartwheels in order to prevent banks, especially big banks, from failing.

Our current economic crisis is clearly a combination of Paths A and B, since no one can remember the last time the US Federal Reserve actually raised interest rates or the last time anyone expressed more than a pro forma concern about inflation.

Conventional wisdom has it that the way to break out of the deflationary spiral is to lower interest rates so as to put more money into circulation. That flow of money is then supposed to jump-start the economy, leading to more jobs, more consumption, and eventually to an increase in prices. But what happens when the deflationary spiral starts at a time when interest rates are already at near rock-bottom levels, which in fact is where they've been since early 2000? Unlike inflationary times when the central bank can raise rates as much as it likes to hold back price increases, rates cannot drop below zero to combat deflation. Often this factor is termed the "liquidity trap." The only way out of it is for the government to pump huge amounts of money into the economy through spending. This is how governments around the world ended the Great Depression of the 1930s.

Today, this spend-till-you-drop path is pretty much closed out as well, due to the massive indebtedness of the United States and the countries of Europe (not to mention the influence of anti-big-government movements such as the Tea Party activists in the United States). To inject the much needed cash into the economy, governments must somehow have that money available. It can come from several sources, each with its own set of associated problems. The obvious first source is lenders like China, Japan, and other Asian countries, who have been sending their burgeoning savings abroad to support an out-of-control lifestyle in the United States and Europe for years. Or it can come from running the printing presses 24/7 to magically make the money appear from paper. Lenders are now very leery about putting good money after bad into the US Treasury. Moreover, the creation of money out of paper opens up the very real possibility of *hyperinflation*. It's difficult to imagine, but this is an even worse solution than enduring a period of deflation as a way of wringing the excesses out of the financial system created by the speculative bubble of the 1990s. Hyperinflation will wipe out the dollar, it will wipe out what remains of the American middle class, and it will eventually wipe out the entire economy. If you don't believe this, have a quick look at Weimar Germany in the early 1920s, or for that matter, Zimbabwe today. Other candidate sources for funding include raising taxes on either individuals or corporations, both politically taboo topics just about everywhere. Moreover, it's difficult to see how taking money out of the pockets of citizens or corporations can help pump up consumer spending, which represents over two-thirds of an economy like that in the United States. Finally, there's the "PIGS solution," being tried today in Portugal, Ireland, Greece, and Spain, which calls for tax hikes *plus* savage cutbacks in government services, ranging from health care to pensions to education.

What's important to keep in mind here is that undoing deflation involves more than just pumping money into the system. The solution is at least as much psychological as it is economic, since the effect of a growing complexity gap often manifests itself in a slowdown in a society's belief that tomorrow will be worse than today (negative mood) leading eventually to a belief that tomorrow will be better, much better than today (positive mood). Once this shift in polarity takes place, people start spending money again because they believe they will either get a job or keep the one they already have. But no amount of government cheerleading or upbeat self-help books and articles is going to bring this about. In fact, the usual way it happens is that some big-league X-event takes place that shocks people out of their funk and into a new psychological orbit. Ominously, this shock is usually a war, a big war—yet one more good reason to pull out all the stops to avoid sinking into the economic depression that is the end point of the deflationary cycle.

For the sake of argument, let's assume that the world of the coming decade or two or three gives us only the second-worst outcome, a global deflation accompanied by a worldwide depression, and manages to avoid the hyperinflation that would tear the world economy apart. What would it be like living in such a world?

Earlier, I made the remark that the word *deflation* is hardly ever uttered in polite company nowadays, and that one big reason for this is that—as mentioned earlier in connection with the seemingly relentless growth of markets—there is no one alive today in the United States who can remember living through such a period. However, there is an entire nation of more than 130 million people who are alive today who can give a very up-to-the-moment account of what it's like living in such a world. Of course, I'm talking about Japan, a country that's been in a deflationary depression for over two decades with no end in sight. In many ways, the Japanese experience since the late 1980s is a kind of dress rehearsal for what the rest of the world can expect to see in the coming years. So it's worth a few paragraphs to detail the "lowlights" of that experience.

BY LATE 1989, THE GROUNDS OF THE EMPEROR'S PALACE IN CENTRAL Tokyo was reputed to be worth more than the entire state of California. Can you imagine? Just a few months later, in early 1990, Japan underwent a similar type of property and stock market bubble burst that the United States and Western Europe experienced in 2007–2008. For example, at its high on December 29, 1989, the Nikkei Stock Index in Tokyo—the Japanese equivalent of the Dow Jones Industrial Average in the USA—stood at 38,876. Now, twenty-two years later, it is less than *one-quarter* of this level. So deflationary bear markets like the one Japan (and soon the rest of the world) is experiencing can take a long, long time to recover (think decades). As a benchmark of comparison, the Dow Jones Industrial Average took twenty-three and a half years to regain its level just prior to the Great Crash of October 1929. So despite massive amounts of monetary inflation going on around the world nowadays, especially in Japan itself, no one thinks the Nikkei will regain its 1989 high anytime in the foreseeable future. The Japanese economy collapsed into a deflationary spiral in early 1990 and has not pulled out of it since.

The situation is not any better for housing prices in Japan either. Currently, the average price of a home sits at the same level it was at in 1983, nearly three decades ago. And while newcomers to the debtor's prison like Greece, Italy, France, and the United States get all the front-page attention nowadays, Japan actually faces the world's largest sovereign debt to other nations, about 200 percent of gross domestic product, a financial burden accompanied by major social problems like an increase in poverty and rising suicide rates.

In his recent book addressing lessons the ongoing Japanese deflation offers to the world, Richard Koo, chief economist at Nomura Securities, makes the following statement about Japan today: "Millions of individuals and companies see their balance sheets going underwater, so they are using their cash to pay down debt instead of borrowing and spending." This decline has been a very corroding experience for the Japanese. In the 1980s, Japanese people were confident, looking forward to the future, and eager to create a new world order in Asia. Today? Well, it's a nation that has lost its self-confidence, fearing a future that its aging, shrinking population is in no position to confront. As a small indicator of this fact, in an article published in the *New York Times* in 2010, Martin Fackler reports a Tokyo apparel shop owner saying, "It's like Japanese have even lost the desire to look good."

A very painful indicator of the effect of a life of deflation and economic stagnation is the attitude of young people toward consumption. Instead of streaming into Akihabara, the high-tech district of Tokyo, to scoop up the latest in electronic gadgetry, many young Japanese refuse to buy any expensive items. As Fackler also noted, a generation of deflation has gone beyond just making people unwilling to spend. It has given rise to a deep pessimism about the future and a fear of risk. Consumers now see it as foolish to buy or borrow, which further accelerates the downward spiral. Hisakazu Matsuda, a keen commentator on this phenomenon, calls Japanese in their twenties "the consumption haters." He says, "These guys think it's stupid to spend." Another observer, Shumpei Takemori, an economist at Keio

University in Tokyo, remarks that "Deflation destroys the risk-taking that capitalist economies need in order to grow. Creative destruction is replaced with what is just destructive destruction."

How did the Japanese government try to pull out of this spiral? No points here for guessing the answer. They did just what Western governments are doing right now. They cut interest rates to zero in 1999 and left them at this literally rock-bottom level for seven years. In addition, the government enacted one bailout scheme after another, together with offering an endless series of stimulus packages. But all to no avail. Added to this is a seemingly potent, but actually impotent, combination of monetary and fiscal policies, along with market and protectionist regulations. To date, nothing has worked. More than two decades after the deflationary spiral started, Japan is still on the edge of total economic collapse. As an indicator of this fact, in early 2010 the Japanese Statistics Bureau reported that prices in Japan have been falling for the past twelve straight months, and that land prices are half what they were twenty years ago. One year later, the situation has hardly improved. In August 2011, Junko Nishioka, chief economist at RBS Securities Japan, noted that "Consumer prices are unlikely to rise much . . . as weak consumer sales are likely to trigger further price competition."

In fairness to Western economies now struggling with the same problem, there are major differences between the Japanese situation and what we see in the United States and Western Europe. The United States can simply print bundles of reserve currency and export it to the rest of the world for products like cars, T-shirts, computers, and other gadgets that will distract people's attention from the real problems the country faces. Moreover, even during the difficult period in Japan, savings grew and the country continued to produce real goods for export. So what to do? The only thing that's certain is what *not* to do: do not, repeat do *not*, continue to pile deficit upon deficit. If there was ever a real-life example of the principle that you can't borrow your way out of a deflation and bring the economy back to life, Japan is that example. Pushing the problem onto future generations can only ultimately lead to an even bigger social crash. On this dynamic, but hardly uplifting note, let's try to summarize the dimensions of the emerging global depression.

ADDING IT ALL UP

THE AUSTRIAN-AMERICAN ECONOMIST JOSEPH SCHUMPETER introduced the term "creative destruction" to describe the process by which the disappearance of outmoded, unnecessary components of an economic system are destroyed to free up space for innovative, new forms of economic production and consumption. We are in the destructive phase of Schumpeter's picture right now, as the global financial and economic systems are transiting from the "old world" of the post–World War II set of structures and rules for economic, political, and social discourse into what will become the standards for the first half of the twenty-first century. The problem of the moment is that no one really knows what that new global structure will be. All that's known for sure is that it will be something very different from the old regime.

Like all dynamical processes, the destruction phase of the Schumpeterian cycle must have an engine driving the process. In this chapter, I've argued that the engine turning the financial and economic worlds upside down is a rapidly approaching period of massive deflation (or, perhaps even worse, hyperinflation). Thus, whatever the picture turns out to be in the longer term (ten to twenty years downstream), there's nothing nice on the immediate horizon. Only when the global system has settled into the creative phase will we reap the benefits of what's to come in the balance of the current century.

PART III

X-EVENTS REDUX

ANATOMY OF AN X-EVENT

AN X-EVENT IS NOT A BLACK OR WHITE AFFAIR. THERE ARE DEGREES of surprise, just as there are degrees of impact. Anticipation of such outliers is also a fuzzy matter. An event in what we've called the "normal" region can smoothly transition into the X-event domain, as one or both of these factors—surprise and impact—passes though a very ill-defined gray zone. Let's try to keep this matter in mind as we go through this concluding part of the book in an effort to say something both meaningful and useful about how to anticipate and prepare for X-events.

Almost the first thing journalism professors do on the opening day of class is put on the chalkboard a list of the six big questions any story must answer: Who? What? Why? When? Where? and How? So it is with X-events, as well. In the preceding parts of the book, I've attempted to tackle all of these, with the notable exception of "When?," which is arguably the most important of the six concerns for anyone who wants to predict, prevent, or limit their exposure to a devastating extreme event. To answer it, I must carve up the landscape of knowledge using some type of taxonomy. So regarding the timing of an X-event, here I'll look from the three vantage points of *Before* the event has taken place, *During* its unfolding, and *After* the impact has been fully experienced and assimilated.

Before: Prior to the event taking place, our focus naturally should be on anticipating the event, determining as accurately as we can when and where it will strike. These are the two areas where science, coupled with imagination and a possibly good database of past occurrences, can be most helpful, as I'll outline in the next section. As I have repeatedly emphasized, though, such a database is the missing ingredient for the overwhelming majority of X-events. (Otherwise, I would not have to have written this book.)

Another vexing question fits into this category as well. Suppose you receive a credible signal of an impending X-event, say unusual seismic activity around a volcano, signaling a strong possibility of an eruption soon. Who do you tell? The answer is more of a sociopolitical matter than it is a question that science can help answer, as there are so many conflicting vested interests at stake in almost any X-event's occurrence that it's impossible to disentangle them from one another. Other than in extreme situations, such as with an asteroid impact or a worldwide pandemic, where the global threat is clear and immediate, human-caused X-events have many nuances and shades of gray behind which every sort of worldview and vested interest can hide. In fact, almost always the interests are ultimately financial. People—investors, corporations, politicians, countries—are making too much money (or receiving it through campaign contributions and taxes) by perhaps riding a stock or housing bubble, so we need to take steps to restore the system to sustainability. But it's damnably tricky to take effective action in the face of short-term thinking on the part of powerful forces whose interests reside in preserving the status quo. As another case in point, witness the heated debates surrounding global warming, where science has identified an emerging dire phenomenon but the political will to do something meaningful about it has not materialized. I can say from personal experience that the very same conflicting interests are present in any kind of early warning of earthquakes, floods, tornadoes, hurricanes, or volcanic eruptions, not to mention advance warnings of human-caused events rather than those dropped in our path by nature.

During: This is the easy part—if you consider surviving an X-event as being easy! When you're in the midst of a food crisis, financial meltdown, or earthquake, you haven't got a lot of time to spend pondering your philosophical navel. Basically, the During phase consists of real-time disaster management, not scientific speculation or worrying and wondering about what went wrong.

After: In the aftermath of the X-event, we enter the "clean up the mess" phase. During this time, we face very practical and immediate concerns of restoring services and facilities destroyed by the event, such as electrical power, communication, housing, food, and water. This period also involves a lot of soul-searching, finger-pointing, postmortems, and patching of the dike to supposedly prevent the event from taking place again. Of course, in practice much of this activity is more like planning to fight the last war than it is in getting ready for the next one. Anyway, this phase again has little to do with the concepts and methodologies of science, futurology, and the like, and almost everything to do with political posturing, coupled with a bewildering array of ingenious and self-serving cover-ups and obfuscations.

The bottom line, then, is that the only phase in which scientific analysis plays a meaningful role in the study of X-events is the first one, Before the event actually takes place. After that, science and planning is literally swept aside by events, and we move into the sociopolitical and psychological spheres where almost anything could happen—and often does. With this in mind, let's return to the question of the type of tools we have or need to develop for getting a handle on when and where the shadow of trouble is going to fall next.

AHEAD OF THE CURVE

Peter and Paul are small, isolated undeveloped lakes in northern Wisconsin. In 2007, they served as the venue for one of the most important ecological experiments in recent times. A team of researchers headed by Steven Carpenter of the University of Wisconsin used these lakes as the site to test the possibility of anticipating radical changes in a system, perhaps even far enough in advance to stave off an environmental catastrophe.

As Carpenter tells the story, "For a long time ecologists thought radical changes couldn't be predicted. But we've now shown they can be foreseen. The early warning is clear. . . . The concept has now been validated in a field experiment and the fact that it worked in this lake opens the door to testing it in rangelands, forests and marine ecosystems." So just what, exactly, did the Carpenter team do? And how did they do it?

Using Paul Lake as the control, the Carpenter team experimentally manipulated Peter Lake by gradually introducing predatory largemouth bass into the lake. Prior to adding these invaders, the lake had been dominated by smaller fish that fed on the lake's water fleas, which served as the zooplankton for the lake ecology. What the scientists were aiming to do was disrupt the lake's food web to the point where it would shift to a system dominated by the predatory bass, pushing the smaller fish down the food chain. In this shift, the researchers expected to see a rapidly cascading change in the lake's ecosystem that would impact all the plants and animals in major ways.

As soon as the predatory largemouth bass were added, the small fish recognized the threat and began to stay away from the open water, confining their exploration for food to areas near the shore and around protective barriers like sunken logs. According to Carpenter, the lake became "water flea heaven," in which the preferred food of the water fleas, the lake's phytoplankton, began to fluctuate wildly. The entire ecosystem then dramatically "flipped" into a new mode of behavior. What the group observed was that computer models mirrored the reality of the ecosystem, as the phytoplankton underwent extreme shifts in its levels just prior to the water flea regime shift.

Note what's going on here. The small fish are now risk averse, and so don't venture into open water to eat so many water fleas. The water flea population then soars, leading the fleas to devour their preferred meal, the lake's phytoplankton. This in turn ultimately leads to a dramatic die-off of the phytoplankton, which cannot support the hugely increased water flea population. As their main food source disappears, the water flea population itself then crashes, allowing the phytoplankton to revive, at which point the cycle repeats itself. But at some point the phytoplankton cannot fully recover any longer, and the lake flips into an entirely new mode in which the food web has been totally reconfigured.

Now the million-dollar question: Could this flip be anticipated on the basis of data collected on the chemical, biological, and physical changes happening in the lake? In particular, do these "blips" in the phytoplankton levels serve as early-warning signs of a mode shift in the ecosystem? The answer was provided through work by William "Buz" Brock, a finance theory expert at the university, who employed tools from an area of dynamical system theory termed "bifurcation theory," to show that the odd blips are indeed a precursor to catastrophic change.

As always in such matters, this experiment offered both good news and bad. The positive report is that the experiment validated a theoretical early-warning signal for collapse of the lake's feeding network. The rapid fluctuations in levels of phytoplankton are indeed a tip-off that "something funny" is going on and that you'd better pay attention. The bad news, though, is that to employ this methodology to identify the early-warning indicator requires a lot of data. This means that you must continuously monitor the lake over an extended period of time, collecting as much information as possible on the lake's biological, chemical, and biotic properties. Carpenter notes that it may not be possible to use this procedure for every ecosystem, but that the price of doing nothing may also be very high.

So here we have a living example of an effective early-warning procedure to identify signals of upcoming catastrophic change: look for wild fluctuations in the behavior of some variables in the system. These rapid blips are advance indicators that the quantities measured (like the phytoplankton in Peter Lake) are entering a danger zone. Ecologists like Carpenter are leading the charge in applying this theoretical methodology to real-world ecosystems.

MATHEMATICAL TOOLS FOR ANTICIPATION

THE EXPERIMENT OUTLINED ABOVE BY CARPENTER & CO. shows clearly that a rapidly increasing rate of fluctuation in Peter Lake's phytoplankton level was a reliable warning sign that the lake was getting ready to flip into a new mode of behavior. This is the first of five early-warning principles that mathematicians have distilled out of the mathematical theory of systems whose behavior changes over the course of time (generally termed "dynamical systems" in the professional literature). Here is the full list.

An increasing rate of fluctuations in the value of one or more of the fundamental properties of the system that you're looking at—such as phytoplankton levels in Peter Lake—is a tip-off that the system is undergoing a major structural change. These fluctuations might turn up as increased volatility of trading volume on a stock exchange, rapid shifts in positions adopted in the rhetoric of political leaders, jittery movement of the ground around a seismic region, or wild changes in the output of an agricultural system. But in all cases, quick swings from one extreme to another often constitute a harbinger of new things to come.

High-amplitude fluctuations are another important signal. The distance the system moves from its high point to the low is the key element to observe here. In other words, it's not just rapid changes (oscillations) in behavior that count, but also whether the behavior is attaining higher highs and sinking to lower lows. Once these peaks and valleys become large enough, the system is often at the breaking point at which one more seemingly minor shove is all it takes to push it over the edge into a totally new mode of activity.

Critical slowing down is the next early-warning principle. Think of a ball placed at the bottom of a mixing bowl having very steep sides. If the ball is moved just a bit, the steepness of the sides will ensure that it rather quickly returns to rest at the bottom. But if you place the same ball in a very shallow-sloped bowl, then it may take quite a while for it to get back to the bottom, as the ball will go back and forth from side to side many times before coming to rest. This latter situation is what a dynamical system theorist would term "critical slowing down," as the system's observed behavior seems to have a hard time recovering from the effect of even a small disturbance. This is a very important early-warning signal that the system is approaching the danger zone where the likelihood shoots way up of a sudden major shift in its behavior.

A network on the edge of a major shift often starts showing a pronounced preference for "visiting" only a small subset of its possible states. In other words, the states that the system actually visits are very unevenly distributed, as the system's trajectory tends to remain in a small subset of all possible states. A system theorist would say there is a pronounced *skewness,* or clustering, in the actually realized states of behavior. Income distribution in the United States today is a good example, in which the distribution is highly skewed to the rich and to the poor, with the midlevels rapidly declining. This is a skewed distribution of incomes (i.e., states). Again, this is an important early-warning signal of an impending X-event, suggesting the rich will not be able to maintain this pronounced distance from the pack for too much longer.

It's worth noting here that this imbalance/skewness of states is very much in the spirit of the complexity mismatch that I've noted many times throughout the book. One system, the rich, have a very high-complexity lifestyle with a huge number of alternative actions they can take at any moment in time (homes to buy, places to travel, food to eat, and so forth). On the other hand, the poor lead a low-complexity life, having few lifestyle choices at their disposal. The gap is widening and will certainly have to be narrowed in the future, either through voluntary action by the rich, government intervention in the process, or what's most likely in my opinion, an X-event along the lines I presented in the story about deflation in Part II.

Many systems change not only in time but also in space. For example, the income gap cited above changes in a very different manner in New York City or Berlin than it does in Nebraska or rural Brazil. Demographic patterns in urban neighborhoods or the density of vegetation in an arid region are good examples of important variables whose values are markedly different not just over time, but from one place to another at the same time. In natural systems like ecosystems or animal populations, *rapid changes in spatial patterns* are often signals of an impending quick mode shift. Many scientific articles cite examples where the climate in a semiarid area gets drier, leading the vegetation to grow in a manner that's much more sparsely and unevenly distributed than it would when all plants get sufficient water. This pattern of patchiness unfolds gradually until it reaches a tipping point, at which all the plants that remain die off and the region turns into a desert. So look for changes in standard patterns as a clue that the system is entering a danger zone.

These dynamical system tools usually require a lot of data for their use. By the very nature of an X-event, like a financial crash, a hurricane, or a political revolution, they don't occur all that often. So as we have already noted many times, it's almost always the case that the data just isn't there. Or at least it's not there in sufficient quantity and quality to effectively use either extreme-events statistics or dynamical system theory for anticipating what's coming up next and when. What to do then? Well, when the real world doesn't provide the data you need, you create a surrogate world that does! This is the idea behind *agent-based simulation,* pseudo-academic language for what amounts to a computer game.

COMPUTER-BASED SIMULATION TOOLS FOR ANTICIPATION

A LONG TIME AGO, BEFORE I FINALLY GREW UP, I HARBORED AN INOR-dinate fondness for betting on games of the US professional football league, the NFL. Being a reasonably analytic sort of guy, I also labored under the same illusion that infects many analytically minded folks about the stock market, namely that there must be some magic formula to massage the data available into consistently reliable forecasts about how the games will turn out. If I lived in bettor's heaven, this formula would enable me to overcome the house odds offered by the bookmakers in Vegas. So I scoured the computer programs on offer at the time, looking for that magic bullet. Needless to say, I am now a lot wiser—but poorer—about the prospects of beating the house at football betting. But I did learn a number of things in that search (and elsewhere) that stood me in good stead thereafter. Let me explain one of them, since it is by far the most relevant lesson and has contributed to my remaining gainfully employed for the past several years.

Most of the programs I experimented with were statistical in nature. In other words, they first collected all the past data as to how many points a team had scored, the number of yards they'd gained running and passing, their advantage in turnovers like fumbles made/recovered, and so forth. The program then processed this data into a statistical estimate of how many points the team would score in its next game. Mathematical modelers would call this a "top-down" approach to the problem, since it totally ignores the individual players and their performances, focusing instead on aggregated measures of how the players performed. This is analogous to looking at a corporate balance sheet rather than digging down into the company's operations and examining how its workers actually produce those aggregated numbers like gross revenues, labor costs, and profits. Such measures may (or may not) be useful in analyzing a company, but I thought they were a rather poor way to analyze how a football team would perform on any given Sunday.

What I sought was a "bottom-up" model that directed attention to the individual players themselves, their playing characteristics like speed, strength, and agility, along with the rules each player used to play his position. With that information, one could put the players into interaction and see what kind of results (points scored) would emerge. Readers will recognize that this approach emphasizes the idea of emergent phenomena, one of the seven pillars of complexity outlined in Part I.

It turned out that there actually existed such a program, and I used it for a few seasons in making my bets. I could actually simulate every Sunday's games inside my computer, play them each a hundred times or so, and look at how many times one team prevailed over the other and by how many points—just the information needed to place my wagers. It's well worth our time to see how this bottom-up approach allowed me to address "What if?" types of questions about any particular game: What if Player A is injured? What if field conditions are wet and muddy? What if . . . ?

This very same principle is at work in "agent-based simulations," where we carve out a piece of the real world and then create scenarios about that slice of reality inside our computers. This amounts to using the computer as a laboratory to do the kinds of controlled, repeatable experiments the scientific method demands but which reality usually will not allow. For instance, you may have a hypothesis about how the financial markets function that you'd like to test. Unfortunately, you can't go down to Wall Street and ask them to change the trading rules in order to test your theory. But you can build a surrogate copy of Wall Street in your computer, populate it with a collection of traders using different rules for trading, and set them into interaction in accordance with the dictates of your theory. If you've done a good job of picking the rules for the traders and even the rules by which they change their trading rules, together with other factors affecting the broader way in which trades take place, then you can expect to get some useful insight into how viable your new theory may be. We can do the same thing in testing hypotheses about the likelihood and possible impact of extreme events.

Before I describe this process, let's be clear on what the components are of an agent-based model/simulation.

- *A Medium-Sized Number of Agents*: The standard jargon term employed to describe the objects composing our system of interest is agent, be it a trader in a financial market, a driver in a road-traffic system, or a country in a geopolitical system. In contrast to simple systems, like superpower conflicts, which tend to involve a small number of interacting agents, or big systems like a galaxy (which has a large enough population of agents—stars, planets, comets, and the like—so that we can use statistical procedures to study them), complex systems involve what we might call a "medium-sized" number of agents. What constitutes "medium" may vary from case to case, but generally it means a number too large for intuition and hand calculation to illuminate the system's behavior and too small for statistical aggregation techniques to provide useful answers to our questions. In a football game, this number is around thirty or so, consisting of the twenty-two players on the field, plus the coaching staffs from both sides. So just as with Goldilock's porridge, a complex system is formed of a number of agents that is not too small and not too large, but just right to create interesting and meaningful patterns of emergent behavior.

- *Intelligent and Adaptive Agents*: Not only are there a medium-sized number of agents, the agents are clever and have the ability to learn and change their behavior as events unfold (adaptive). This means that they make decisions on the basis of rules, like the principles a quarterback or a linebacker uses to play his position in football. Moreover, the agents are ready to modify their rules on the basis of new information that comes their way during the course of play. In some cases, the agents can even generate new rules that have never been used before, rather than being hemmed in by having to choose from a set of predefined choices for action. For example, the great San Francisco 49er receiver R. C. Owens introduced the "Alley Oop" play into his repertoire of ways to catch the ball over the outstretched arms of a defender. This adaptive aspect, in fact, is what tends to separate the great football players (like Peyton Manning, who seemingly changes his offense's play before every snap based on the defense's alignment) from the journeymen. This ability to generate new rules means that an "ecology" of rules emerges, one that continues to evolve as the game, and even the entire season, unfolds.

- *Local Information*: No one player has access to what *all* the others are doing at a given stage of play. At most, each player gets information from a relatively small subset of the other players and then processes this "local" information to come to a decision as to how to act (i.e., what rule to use). In most complex systems, the agents are like the drivers in a road-traffic network or traders in a speculative market, each of whom has information about what at least some of the other drivers or traders are doing—but not all of them.

You might well argue that while a football game is an interesting little puzzle, it is of no great import when it comes to X-events, even at the very restricted level of football fans and bettors like my former self. While it's definitely possible to use "Football World" to do many experiments, and even explore different situations that may well give rise to surprises, the impact factor is definitely missing that would turn any such surprise into an X-event. But it's the idea of creating a world in the computer to generate data about a system in which bona fide X-events may arise that is the real message here. So let's examine another computer simulation example that I myself was recently engaged in that has all the features for the appearance of a true X-event.

• • •

IN MY DAY JOB, I HEAD A RESEARCH PROJECT DEVOTED TO TRACING the potential impacts various X-events may have on the economy of countries. An agent-based simulation of the global trade network created along the lines we've just discussed is a key tool in this analysis. Let me give just a brief summary of one geopolitical scenario we studied recently. The X-event in question started with the question: What would happen if China were to begin to overtly challenge US global hegemony? This challenge might come in three different forms, leading to the following different scenarios for exploring the impact on the economic health of the countries in our computer world:

Scenario I: China is increasingly assertive, belligerent even, in all areas of conflict and contention. We might call this the "hard" path. In this scenario, the wildest card of all is a live military confrontation between China and the United States.
Scenario II: China is still assertive but exercises it aggressiveness in a more subtle way, often through diplomatic channels, denial of resource exports, and the like (the "soft" path).
Scenario III: China is weakened due to internal stresses of an economic, political, and social nature, while the United States re-emerges somewhat miraculously as the dominant global power. This scenario involves progressively slower Chinese growth, while seeing the United States regain confidence and influence.

The global trade world in our simulation for investigating these scenarios consists of the following twenty-two countries (agents):

Euro Zone: Finland, Sweden, Denmark, Belgium, Holland, Germany, France, Spain, Italy
Americas: USA, Mexico, Canada, Brazil
Asia: China, India, Japan, Indonesia
Others: UK, Norway, Russia, Turkey, South Africa

Each country has a set of actions it can take at any moment, such as a combination of imposing tariffs, offering "favorable nation" discounts, revaluing its currency, and other such large-scale macroeconomic actions. Of course, these decisions are constrained by both geographical and political considerations, such as trade alliances, costs of transporting goods, and other matters of this sort.

In the first scenario, where China takes the hard path and openly confronts the United States, the big loser by 2030 turns out to be China itself. Interestingly, the countries that suffer least are those in the Benelux nations. But all twenty-two nations are worse off than they were at the beginning in 2010.

By way of contrast, when following the soft path of Scenario II, all countries are better off by 2030 than today, with China leading that charge, achieving a 9 percent higher GDP over the 2010 baseline. The countries that benefit the least under the soft path are exactly those that suffered the least under the hard path.

Finally, the extreme case. Here China collapses from internal stresses while the United States is resurgent. The victors are the North American contingent—but with Mexico being the biggest winner of all, not the United States. Not surprisingly, the biggest loser in this world of 2030 is China followed by Japan, India, Russia, and Brazil—in short, the BRIC countries plus Japan.

I think this is enough to provide the overall drift of how agent-based simulations can be employed to give a forewarning of things to come. This exercise shows that the simulation can help in anticipating an X-event that hasn't yet occurred. It can also shed light on the impact of an X-event that the investigator may drop into the world by, in effect, "playing God."

Throughout the course of this book, complexity—and particularly the idea of a complexity mismatch as the root cause of X-events—has been the thread connecting the many examples and principles I've discussed. Sometimes the thread has been visible on the surface, as with the discussion of the Arab spring in the book's preamble; sometimes it's been a deep undercurrent, as in several of the examples chronicled in Part II. But always the complexity of one system is pitted against that of another, generating stresses that are ultimately relieved by an X-event. It's time now to bring this theme to the foreground again and view it within the light of our earlier deliberations on early-warning signals for an impending X-event. In particular, I want to examine ways of characterizing and measuring the complexity gap between two (or more) systems, and how to use that measurement to foretell, and perhaps forestall, an impending flip from one type of behavior to another.

MIND THE GAP

IN 2011, FRENCH AIR-SAFETY INVESTIGATORS RELEASED THEIR AC-count of the final minutes of Air France 447 as it plunged into the sea off the coast of Brazil late in the evening of June 1, 2009. After a truly heroic search for the plane's two flight recorders, divers eventually found them nearly thirteen thousand feet beneath the Atlantic just a few days before authorities were preparing to abandon the search. An even greater miracle was that the data on the recorders was still intact, allowing the air-safety scientists to reconstruct what was going on, both with the plane itself and the crew in the cockpit during those final, fateful minutes before the plane plunged into the water. The story is as instructive as it is frightening.

As the recorders documented, the plane's airspeed sensors failed, giving the pilots sharply reduced speed readings following the air-craft's entry into stormy ice clouds more than thirty-five thousand feet above the sea. As the speed fell off, the plane entered a stall, wherein the air flowing over the wings was too little and too slow to generate the lift needed to keep it aloft. The stall warning alarms sounded three times, which should have caused the pilots to push the plane's nose down in order to speed up and generate the lift needed to get the airplane soaring upward again. But for reasons still hard to explain, the pilots pushed the nose *up* rather than down, further contributing to the stall. At this point the airplane was no longer flying but falling—at a speed of 180 feet per second (over 120 miles per hour). Just over three minutes later it crashed, killing all 228 people aboard. The biggest puzzle seems to be why the pilots raised the nose instead of dropping it, exactly opposite to the action needed to pull out of a stall. How/why did this happen?

The data from the cockpit voice recorder suggests the pilots may have thought they were taking appropriate action as the speed readings were flipping all over the place. Moreover, the plane was flying over the ocean through a region of moderate turbulence and it was dark, preventing the pilots from being able to see the horizon or any other landmarks that might provide them with a sense of the aircraft's speed, orientation, or direction. As Richard Healing, a former US National Transportation Safety Board member, stated, "All we know is that the information wasn't reliable, and that a lot of warnings were going off and it was probably very, very confusing." Another knowledgeable observer, Bill Waldcock, a professor at Emory-Riddle Aeronautical University, went on to say that "The only thing that would make any kind of sense is that they'd gotten spatially disoriented, they don't know what way is up and they don't fully understand what the airplane is doing." The strangest comment of all comes from the chief French investigator, Alain Bouillard, who said that "They [the pilots] hear the stall alarm but show no signs of having recognized it. At no point is the word 'stall' ever mentioned."

From an X-event perspective, the crash of AF447 is a textbook example of complexity mismatch in action. We have the instrumentation and warning system of the plane giving a variety of both visual and audio signals, which are being read and interpreted by three pilots, only one of whom has actual control of the plane (the least experienced of the three, as it turned out). So the plethora of signals coming from the airplane, coupled with the other signals (or lack thereof) from the environment, seem to have completely overwhelmed the ability of the pilots to sort through all the inputs and come to the correct course of action to restore lift to the plane. In short, the complexity of the system (the plane and its environment) became too great for the complexity of the controller (the pilots) to manage, leading to the crash (the X-event) and the death of all 228 people aboard.

There is plenty of complexity-theoretic meat in this example to keep one busy for days. But the gist of the situation for an X-event researcher is to ask how that complexity mismatch could have been avoided? Or, if not avoided, how could the system be designed to narrow the gap as quickly and reliably as possible, once it did emerge? It's clear that there is much to ponder on both sides of this equation. The plane's instruments and warning system contributed a lot more heat than light to the situation, exacerbating an emergency that already had a very tight time window for action. On the other side, the pilots seemed to be in a state of information overload that prevented them from quickly coming to a consensus on what to do and getting it done in a timely fashion. In short, a combination of confusing, conflicting information processed erroneously, probably coupled with some degree of panic, sealed the fate of the plane and its passengers.

Before entering into the issue of how to recognize a complexity gap and estimate its size, it's useful to spend a few pages discussing how small gaps can be amplified by the technological structures underpinning modern social systems, aided and abetted by human nature itself.

IN THE VOLUME *A DEMON OF OUR OWN DESIGN,* WHICH IN MY VIEW IS certainly one of the most enlightening and accessible accounts of the 2007 financial crisis, Richard Bookstaber, a finance veteran who is now an adviser to the SEC and serves on the Financial Stability Oversight Board, makes the useful distinction between a "normal accident" and what he calls "accidents waiting to happen," those stemming from the complexity and tight coupling of many systems. In the former category are events that can be *expected* to happen, unavoidable even, given the interconnection of various subsystems that compose the overall structure.

The normal accident at the nuclear power plant at Three Mile Island in 1979 is a good illustration of this sort of problem, in which a warning light for a relief valve failed, causing workers to ignore a blocked water valve that ultimately led to the failure of the plant itself. Although accounts of the Three Mile Island accident called it an "incredible" event, the only thing incredible about it was the unbelievably long chain of processes that all had to work *correctly* in order for the plant itself to function. While the likelihood of failure of any one of these processes was minuscule, the odds of at least one process failing was not low at all. A crucial feature of normal accidents is that not only do they arise from too much complexity in the system, in the sense of there being too many interacting parts that must function properly for the overall system to work, but also that the addition of safety checks aimed at combating failures often contributes to the complexity and thus can actually work against the reliability of the system instead of enhancing it.

On the other hand, there are systems whose many parts interact in ways that create behaviors that are counterintuitive, unexpected, or just plain hard to understand. In short, the behaviors are surprising. The global financial markets are a case in point. Earlier, we argued that the very diversity of financial instruments characterize the complexity of the system. But that's only part of the story. The way some of these instruments are structured itself gives rise to surprising behavior. Bookstaber cites the example of a low-cost instrument marketed by Bankers Trust to its customers allowing them to hedge against interest-rate changes. But it slipped past the buyers that the low cost hid a feature of the contract that would cause losses to skyrocket if interest rates moved up too fast. Some buyers of this derivative contract took losses measured in the hundreds of millions of dollars to learn about this feature of their so-called "hedge." A major source of the surprise in these types of situations is that there is generally too little time available to take remedial action before the system spins out of control letting the X-event loose.

In his book, Bookstaber offers the hub-and-spoke network used by the airlines to route flights as an example of a system that is composed of many parts that may interact in sometimes mysterious ways, yet are not tightly coupled. So while it may be annoying to find your flight from Chicago to Albuquerque stuck on the ground because of a thunderstorm in Minneapolis, you have plenty of time to investigate alternate routes to New Mexico while you cool your heels at O'Hare. The system has enough slack built into it to enable it to continue to perform its function of getting you from where you are to where you want to be, albeit with some delay and/or added expense along the way. The point is that the loose coupling prevents a total air transport breakdown.

Generally speaking, the best solution for solving a complexity mismatch is to simplify the system that's too complex rather than "complexify" the simpler system. So, for example, in the case of the financial markets, it would be vastly preferable to eliminate, or at least greatly reduce, the availability of exotic financial instruments that no one really understands rather than to try to beef up the regulatory rules and regulations to control them. The offense always has the advantage, with the defense forever having to play catch-up. Far better to hobble the offense, at least insofar as the overall survivability of the financial system is concerned.

Interacting subsystems and tight coupling make a fiery marriage. And the union is further stressed by the personalities of the parties involved. Specifically, there's the human factor in which people and institutions fail to foresee clear signs of trouble. When it comes to human-caused/induced X-events, it's essential to examine how the foibles of human nature lead us to catastrophically mismanage complexity. A central element is the problem of "not seeing what we don't want to see," or as it's called in legal circles, "willful blindness." Recently, award-winning businesswoman, author, and playwright Margaret Heffernan has studied the phenomenon and published her conclusions in a volume of the same title. Here are a few illustrations of how human nature is conditioned not to see what it should be seeing.

Of special concern for my argument that complexity "bloat" is the root cause of X-events is the case of information overload. Heffernan talks about how multitasking and an overdose of sensory stimulation, taken together with physical exhaustion, can and does narrow the focus of what we see and don't see. Her argument, along with that of many others, including Nicholas Carr, who calls it "cognitive load," is a simple one: it's harder to concentrate when we're tired, since the brain is using so much energy to stay alert that our higher-level brain functions shut down to conserve energy. This, in turn, narrows the focus of what we can and cannot "see." As an illustration, Heffernan tells about an explosion at a BP oil refinery in Texas City, Texas, in 2005. She examined this accident and found that the plant had experienced several rounds of cost-cutting layoffs, forcing the remaining operators to have to work long, tiring shifts, reducing their ability to see warning signals of the impending disaster, which killed fifteen people.

As another illustration of the same process, Heffernan cites the case of Staff Sergeant Ivan Frederick II who was sentenced to prison in 2004 for abusing prisoners at the infamous Abu Ghraib facility in Baghdad. Frederick had been working twelve-hour shifts, seven days a week with very few days off over an extended period. This work regime led to physical exhaustion that was only exacerbated by his being surrounded by colleagues in the same situation. As Ms. Heffernan describes it, "no one was awake enough to have any moral sensibility left."

The herding mentality I've mentioned more than a few times in my story also comes into play in the context of willful blindness, here going under the label of the "Cassandra effect." According to Greek mythology, Cassandra was given the gift of prophecy. But it was coupled with the curse that no one would believe her. Such seers are generally punished heavily in the court of public opinion, which is more than adequate cause for prophets to keep their prophecies to themselves. Again in the context of the Abu Ghraib prison, we have the case of Joe Darby, who gave photos of the abuse of prisoners to his superiors. As Darby told it, "I had to make the choice between what I knew was morally right and my loyalties to other soldiers. I couldn't have it both ways." How he ended up having it is that he was forced to relocate and assume a new identity because some residents in his own hometown regard him as a traitor. So there's often no thanks or honor for being a prophet.

Heffernan's book concludes with the message: "We make ourselves powerless when we choose not to know." In the context of this text, I could paraphrase this by saying we open ourselves up to potentially devastating events when we choose not to see instead of facing reality head-on.

IN A SCATHING ATTACK ON THE GROWING INCOME INEQUALITY IN America, Nobel Prize–winning economist Joseph Stiglitz made the following statement in 2011 (published in the magazine *Vanity Fair*): "The top 1 percent have the best houses, the best educations, the best doctors and the best lifestyles, but there is one thing that money doesn't seem to have bought: an understanding that their fate is bound up with how the other 99 percent live." From this observation, one might draw the inference that the number of lifestyle choices available, ranging from houses, doctors, travel, and education serves as a reasonable measure of the complexity level of a person's life. Put compactly, the more choices you have, the more complex is your life. As a first cut, this is a pretty good measure. If I have the choice of working at a job or not, traveling to Costa Rica or New Zealand for holiday or trekking through the Andes, having my cancer treated at Sloan-Kettering in New York or the Mayo Clinic, or paying for my son to attend an Ivy League school or Stanford, then my life is vastly more complex than if I do not have these choices. In other words, the rich have many degrees of freedom (i.e., choice of action). What I actually choose to do doesn't affect the complexity level of my life, at all. To make it complex, it's enough just to have those choices available. And, for better or for worse, to a large extent it's money that buys those choices in our world.

With this concept of complexity in mind, there's not a shadow of doubt that the USA is experiencing a societal complexity imbalance that's grown exponentially over the past few decades. The gap now sits at a level at which the top 1 percent of the population accounts for over 25 percent of the income, and what's even worse, a whopping 40 percent of total assets. By way of comparison, those two figures stood at 12 percent and 33 percent, respectively, in 1985. By the arguments I've made throughout the book, especially in the stories told about some of the X-events in Part II, a huge awakening is on tap that will result in a rapid and painful X-event to close this gap.

Karl Marx famously stated, "History repeats itself, first as tragedy, second as farce." In these pages I've painted a dire picture of how human-created complexity fits the first Marxian view of history to a tee. This is especially sad since at the same time we are living in the most technologically advanced society ever known to humankind. Yet we continue to sow the seeds of our own destruction, seeds that are for the first time in history able to develop into the destruction of our entire species. This is a good point to start thinking seriously if we will ever reach the second phase of history and be able to look back at our tragic phase with both humor and wonder. I'd like to spend the final pages of this book outlining a few ideas on what we might do today, tomorrow, and the day after tomorrow to redress, or at least minimize, the effects of the complexity imbalances. I believe that there is still plenty of room for optimism—even in a world riddled with X-events.

THE HEDGER'S DILEMMA

THE MOST DRAMATIC X-EVENTS ARE THE ONES THAT GET THE HEAD-lines. Reading about the Bangkok flood in 2011 or thinking about Hurricane Katrina and the levee collapses in New Orleans in 2005, it's easy to become fatalistic about these sorts of natural disasters. And who could blame you? If your thoughts pass from these "minor" natural disasters to something like the asteroid impact or supervolcano eruption I discussed briefly in Part I, it's difficult to be optimistic about your future in the aftermath of such an X-event. They'd be enough to make anyone simply cast their fate to the winds. But such an attitude need not and should not blind us to the fact that the human-caused X-events taken up in Part II are, for the most part, avoidable. Or in the worst case, their damage can be greatly reduced by human attention and preventive action. The many dissections of the 2007 financial crisis illustrate the possibility of X-event avoidance, provided we are willing to change our views as to how such a system could and should function. But changing beliefs, as opposed to feelings, is a painful, difficult, and time-consuming process.

Those of us living in the modern, industrialized world don't yet seem ready to embrace the notion that life is not without risks. We've been coddled and protected to the extent that we actually expect our governments and other public institutions to solve all problems and address our hopes and needs without cost or risk to ourselves. In short, we've fallen into the misguided belief that everyone can be above average, that it's everyone's birthright to live a happy, risk-free life, and that any misfortune or bad judgment or just plain bad luck should be laid at someone else's doorstep. So the first step on the road to reality is to drop these delusions of Utopia. While T. S. Eliot's dictum cited about the human inability to bear too much reality appears on the mark, too much reality and *some* reality are very different matters. Let me illustrate the value of reality over myth by citing work by Monica Schoch-Spana that separates the two.

In a seminar on "Homeland Security, the Environment and the Public" held in 2005, Ms. Schoch-Spana outlined five disaster myths and the associated realities, with special focus on the problem of social behavior following an X-event. Her arguments point to the fact that people's "commonsense" notions about what the public response will be to an X-event—notions based simply upon hunches, intuition, and beliefs—just don't stand up to the facts. Here's one of the examples she presents.

Myth: When life and limb are threatened on a mass scale, people panic. They revert to their savage nature, and social norms readily break down.

Fact: Study after study has shown that in an emergency situation, people seldom revert to a lifeboat mentality, putting themselves first. Most revealing is that on surveys about how people *think* they will behave when disaster strikes, respondents generally say they actually will revert to a jungle mentality. But, in fact, panic is the exception. Instead, creative problem solving is the order of the day.

As an example, the paper cites the 1989 Loma Prieta earthquake in California, where forty-nine of the fifty people pulled out of the rubble were not saved by professional rescuers but by a group of eight Mexican construction workers who happened to be in the immediate vicinity. Similar stories are told of the altruistic responses of those directly affected by the 9/11 and London Underground bombings. I can't resist recounting just one more of Schoch-Spana's myths, since it's so directly relevant to the X-events I've outlined in this book.

Myth: Acts of God and nature are preordained. There is no real way to thwart their ultimate outcome. The same goes for bureaucratic red tape, another so-called immutable force.

Fact: For the period 1975 to 1994, hurricanes were the second-costliest natural hazard in terms of property losses and third most injurious. Improved forecasting and more stringent building codes have now relegated hurricanes to only the seventh-leading cause of death due to natural disasters.

The most important lesson to be taken home from this hugely insightful study is that the outcome of a human-caused X-event is not cast in concrete or inevitable. Human action can dramatically affect the number of lives and/or dollars lost. Moreover, there is often a silver lining to even the blackest X-event-created cloud. To illustrate this point, let's have a quick look, yet once again, at the March 2011

Japanese earthquake and a few of the potential aftereffects on Japanese society.

SHORTLY AFTER THE JAPANESE QUAKE, FINANCIAL COLUMNIST William Pesek pointed out that earthquakes have played a major role in the Japanese psyche as a generator of not only physical trauma but also social change. He cites the 1855 earthquake that not only leveled what is modern Tokyo but also ended Japanese isolation of the Tokugawa period. Reconstruction following the 1923 quake quickly led to the rise of Japanese militarism, while the 1995 Kobe earthquake ushered in the end of the postwar Japanese industrial boom and the subsequent deflation that's been under way in Japan ever since. So, asks Pesek, is historical change again on the horizon?

Here are three possible shifts identified by Pesek that the earthquake might catalyze to help bring Japan out of its decades-long deflationary torpor:

- *Political Shock*: Despite years of deflation and stagnating wages, Japanese officials have dithered over decisive action to address their economic problems, even after China swept past it as the world's second-largest economy in 2010. The earthquake may serve to blast the Japanese government out of its complacency and paralysis. There is no choice now but to rebuild the country and to do so without huge borrowing, but with focus on building up internal economic structures emphasizing entrepreneurial activities.
- *Improved Ties with China*: China's condolences and immediate offer of help to Japan following the quake may go far toward reducing tensions between the two nations. Long-standing disputes over territory, military activities, and the like may be swept away in a new era of "good feelings" between the two countries as a result of the Japanese tragedy.

- *Increased Japanese Confidence*: People around the world have been amazed at Japan's quick response to the quake damage and the orderliness of the Japanese people in the face of such a catastrophe. The lack of looting and social unrest showed Japan to be a stable and caring society, highly civilized, and, in short, a role model of how a society can react to a huge disaster.

The take-home message from this not-so-little story is that an X-event can serve as an opportunity as well as a problem.

Looking at the list of potential X-events considered in Part II, we see a world in which

- oil is running out,
- power grids are stressed to the breaking point,
- the Internet is on the edge of failure, and
- food prices are escalating beyond most people's ability to pay.

This list is pretty intimidating and certainly doesn't inspire an upbeat, optimistic view of the future. But for the most part, the items on this "hit list" are as yet mere possibilities, not yet realities. And even though some of them are rather likely to take place, the good news is that most of them can be anticipated (as I have done here), and some can even be averted. The bad news, though, is that it's difficult to pay attention to mere possibilities, especially when they are rare and the time frame is indefinite. According to what I like to call the Ugly Swan Paradox, even though everyone agrees that surprises always occur, no *specific* surprise ever takes place. This attitude has to be fought at every turn. Specific surprises, even those outlined in Part II, can and do happen. And they happen even if you think they are so unpalatable that you don't want to think about them, and thus fool yourself into believing they cannot occur. They will happen anyway. The damage will be orders of magnitude greater than need be the case by sticking your head in the sand and pretending otherwise.

The distilled essence of my message in this book is that complexity overload is the precipitating cause of X-events. That overload may show up as unmanageable stress or pressure in a single system, be it a society, a corporation, or even an individual. The X-event that reduces the pressure then ranges from a societal collapse to a corporate bankruptcy to a nervous breakdown. But more often the overload comes in the form of two or more systems in interaction, in which the complexity of one of the systems outpaces that of the other(s), in which case a gap emerges. As this gap widens over the course of time, what we might term an "interactive pressure" builds up. If the pressure is not released by a gradual closing of the gap, then it ultimately gets released by a "snap-back" closure in the form of an X-event. So without a social equivalent of a "pressure release valve" to slowly reduce the complexity gap, an X-event lies in the system's future.

Looked at from this perspective, the size of the complexity overload/gap is a new way to measure the risk of an X-event, what we might call the "X-risk." When the magnitude of the overload is very high, the risk is great; when it's low, the risk is less. Our goal both as individuals and as members of society is to act so as to reduce the X-risk. How to do that?

First, note that the problem of reducing a complexity gap comes in two very different flavors. By far, the most important is how to avoid having the gap arise in the first place. As the saying goes, an ounce of prevention is worth a pound of cure. And never was that more true than in this situation. Avoiding the gap means designing our systems so that they work as a unified whole rather than as a collection of systems managed in isolation. The power grid is connected to the Internet which in turn is connected to the financial system and so forth. We can no longer afford to let fat tails wag the societal dog and allow some systems to build up a level of complexity totally out of harmony with the other systems that it both feeds and relies upon.

In times when people perceive X-risk as low, the emotion of greed generally dominates feelings of fear. We saw that in spades during the recent financial crisis when bankers, traders, and investors all bought in to the idea of "free" money in assets like mortgage-backed insurance that seemed too good to be true; it was! So these are the times when regulators should pay special attention to reining in bouts of "irrational exuberance," to use a now discredited catchphrase.

On the other hand, in times when the perception of X-risk is high, fear dominates greed and reduction in complexity is the name of the game. Again, in the current economic climate that reduction has taken the form of closing down subsidiary firms, laying off workers, and the like. Such actions also need to be carefully monitored and even regulated to ensure that they don't throw out the baby with the bathwater. Too little complexity in one part of the overall system without corresponding reductions elsewhere leaves the very same X-event-generating gap. There's no free lunch. You must add and subtract complexity judiciously throughout the entire system in order to bring the imbalances back into line. Focusing on just one or two subsystems like finance or communication while ignoring the others will give only the illusion of progress, an illusion that an X-event will soon shatter. Much the same line of argument applies to reducing an existing complexity gap as opposed to avoiding it. Here are a few general principles that apply in either case.

First and foremost, systems and individuals should be as *adaptive* as possible. The future is always an unknown and scary place. Today it's even scarier and more unknown than usual. So developing yourself and our infrastructures so that they have more degrees of freedom to counteract or exploit whatever may come their way is a good basic strategy.

Closely allied to adaptivity is *resilience*. This can combat complexity overload by creating an overall infrastructure that is capable of rolling with the punches. In fact, not just rolling but actually benefiting from the punches. A good example comes from the forest industry. Forest managers regularly set *controlled* fires to burn off dead wood and overgrowth that would otherwise serve to fuel major, uncontrollable fires when lightning, errant campers, arsonists, or other unforeseeable events inevitably set the forest ablaze.

Redundancy, building in spare capacity, is a tried-and-true method to keep a system running in the face of unknown and often unknowable shocks. Every computer administrator knows this, and so does most everyone else who uses a computer. The point is to have plenty of spares ready to bring online when one or another component of the system fails. This X-risk minimizing tool is closely allied to the No Free Lunch complexity principle presented in Part I. How do we balance out the cost of maintaining a robust system against the loss of economic efficiency that the backup entails? At the personal level, it costs money and time to purchase the software and hardware to create and store regular backups of the data on your computer or external hard drive. It also costs time to actually carry out those operations. But if you're the author of a book like the one currently in your hand, you will be very, very happy to pay that economic price in exchange for the peace of mind in knowing that a year's worth of work can't be wiped away by a false keystroke or a power surge.

Of course, adaptation, resilience, and redundancy are just general principles, guidelines if you like, not a detailed plan for either preventing or combating the complexity overload that underlies X-events. The principles must be interpreted within a given context, both as to what they actually *mean* in that context, along with how that meaning can be translated into actions that apply the principle within the given setting. And this is the case whether the setting is your personal life or the life of an entire country or even the world. Complexity overload is not inevitable. But it is endemic. And as with the price of freedom, eternal vigilance is also the cost of keeping X-events at bay.

My final word, then, is to accept that X-events will occur. That is simply a fact of life. So prepare for them as you'd prepare for any other life-changing, but inherently unpredictable, event. This means remaining adaptive and open to new possibilities, creating a life with as many degrees of freedom in it as possible by educating yourself to be as self-sufficient as you can, and not letting hope be replaced by fear and despair. Humankind has survived X-events much worse than those on our list in this book and will do so again. Cartoonist Walt Kelly's character Pogo stated, "We have met the enemy and he is us." The more we can do to change this dictum, the better we'll be able to weather whatever comes our way.

NOTES AND REFERENCES

Preamble: Putting the "X" into X-Events

An enlightening potpourri of books addressing the extreme events and the concomitant social problems at various levels of academic sophistication and at various levels of details include the following:

> Warsh, D., *The Idea of Economic Complexity* (New York: Viking, 1984).
> Posner, R., *Catastrophe: Risk and Response* (New York: Oxford University Press, 2004).
> Clarke, L., *Worst Cases: Terror and Catastrophe in the Popular Imagination* (Chicago: University of Chicago Press, 2006).
> Rees, M., *Our Final Hour* (New York: Basic Books, 2003).
> Leslie, J., *The End of the World* (London: Routledge, 1996).
> Homer-Dixon, T., *The Upside of Down* (Washington, DC: Island Press, 2006).

This collection is a great introduction to the theme of this book. The book by Warsh is one of those rare volumes that in my view will be seen in the historical perspective as the forerunner of an entirely new way of looking at economic processes in particular, and social processes in general. Richard Posner is a federal judge in Chicago and presents his litany of catastrophes in a calm, reasoned, concise, almost antiseptic fashion. The book by Lee Clarke is as much about the psychology of the potential victims of terrorist attacks and natural calamities as it is about the events themselves, and thus makes a good counterpoint to Posner's detached, analytical discussion. Martin Rees is one of Britain's most distinguished scientists, former president of the Royal Society and England's Astronomer Royal. His book is written for the curious layperson and, naturally enough, emphasizes nature's ways of doing us in. Leslie is a philosopher by profession and brings a philosopher's mind to the logical analysis of whether the human race is in imminent threat of extinction, concluding that we probably are. His treatment is both academically thorough and easy to read—a rare combination. While the foregoing volumes are a bit on the gloomy side, Homer-Dixon's account of the predicament we're in today offers hope for a way out of our current dire situation. His book is a tour de force on how to make our society resilient enough to survive into the next century.

1 If you're interested in seeing **Bryan Berg** and his huge house of cards, check out

the following website for details: http://newslite.tv/2010/03/11/man-builds-the-worlds-largest.html.

6   For the details of the baseball simulation showing that **Joe DiMaggio's hitting streak** was not so special after all, see Arbesman, S., and S. Strogatz. "A Journey to Baseball's Alternate Universe," *New York Times,* March 30, 2008.

8   The **analytical formula** mentioned in the text for characterizing the "X-ness" of an X-event is $X = IM(1 - UT/(UT + IT))$, where IM is the impact magnitude measured in normalized units, such as dollars of damage versus total GDP or lives lost versus total annual deaths, in order to ensure that IM is a number between 0 and 1. If you don't care about this normalization, then using absolute deaths or dollars is fine; the final result will still give a sense of the relative extremeness of the event, it just won't be a number between 0 and 1. The quantity UT is the unfolding time of the event, while IT is the event's impact time. The final value of X is then a number between 0 and 1; the larger this value, the greater the "extremeness" of the event. Just to be clear on the matter, I do not take this formula very seriously as a precise measure of the magnitude of an extreme event; it's simply a guideline, or rule of thumb, for comparing such events.

10   An interesting blog item on the problem of **complexity collapse and modern society** is given in the following post by former US Army Intelligence officer James Wesley Rawles, who published the recent novel *Survivors,* outlining how society might look when all infrastructures we rely upon for daily life disappear overnight: http://www.survivalblog.com/2010/06/is_modern_society_doomed_to_co.html.

12   The original account of the **law of requisite complexity** was given by cyberneticist W. Ross Ashby in 1956, who called it the law of requisite variety. Perhaps this is a better name anyway, as it suggests the notion I've emphasized in this book of complexity being tied up with the idea of diversity of actions, the degrees of freedom, that a system has at its disposal to address whatever problem may come along. Ashby's exploration of this idea is given in his pioneering book *An Introduction to Cybernetics* (London: Chapman and Hall, 1956).

A recent account of the basic idea in the context of the world of commerce is given by international business consultant Alexander Athanassoulas in: Athanassoulas, A., "The Law of Requisite Variety," *Business Partners,* January–February 2011, 16.

Part I: Why Normal Isn't So "Normal" Anymore

20   A good summary of Ambrose's work on nailing down the evolutionary bottleneck that the **Toba volcano** created is available at the website: www.bradshawfoundation.com/stanley_ambrose.php.htm. The full details are given in: Ambrose, S., "Late Pleistocene Human Population Bottlenecks, Volcanic Winter, and Differentiation of Modern Humans," *Journal of Human Evolution, 34* (1998), 623–651.

24   The concept of what I'm calling **"complexity overload"** has been in the air for several years now. Here is an eclectic sample of some ideas that have been put into circulation on the Internet that identify and explore this concept to help understand financial crises, the Internet, the Arab spring, and plain everyday life:

Helgesen, V., "The Butterfly and the Arab Spring." Editorial in *International IDEA* (www.idea.int/news/butterfly-arab-spring.html).

Barratt, P., "Systemic Complexity, the Internet, and Foreign Policy" (http://belshaw. blogspot.com/2010/12/).

Nickerson, N., "On Markets and Complexity," *Technology Review,* April 2, 2011 (www.techologyreview.com).

Danielsson, J., "Complexity Kills" (www.voxeu.org).

Norman, D., "The Complexity of Everyday Life" (www.jnd.org/dn.mss/ the-complexity_of_ everyday_life.html).

30 The **mousepox near disaster** is described in "The Mousepox Experience," *EMBO Reports* (2010) 11, 18–24. (Published online: December 11, 2009.)

34 The statement from **General Carl Strock** cited in the text was taken from the following interview with Margaret Warner on PBS: http://www.pbs.org/newshour/ bb/weather/july-dec05/ strock_9-2.html.

37 Nassim Taleb's best-selling volume bringing the reality of **fat-tailed distributions** to the attention of the general public is: Taleb, N., *The Black Swan* (New York: Random House, 2007).

42 The statement by **Ray Ozzie** about the suffocating effect of complexity was quoted in the following article: Lohr, S., and J. Markoff, "Windows Is So Slow, but Why?," *New York Times,* March 27, 2006.

44 The popular account of societal collapse put forth by Jared Diamond in the book cited below is the version that's attracted attention in recent years. But the earlier work by **Joseph Tainter** is the one that will warm the heart of every complexity scientist. In any case, they're both fantastic reads:

Diamond, J., *Collapse* (New York: Penguin, 2005).

Tainter, J., *The Collapse of Complex Societies* (Cambridge: Cambridge University Press, 1988).

A stimulating popular account of the arguments in both these books can be found in the article: MacKenzie, D., "Are We Doomed?," *New Scientist,* April 5, 2008, 33–35.

47 In the two decades or so since the **Santa Fe Institute** popularized the notion of complexity and complex systems, many SFI alums and others have put out popular accounts of this developing paradigm (including yours truly). Here are a few entrées for the interested reader to get a feel for the subject:

Casti, J., *Complexification* (New York: HarperCollins, 1994).

Cowan, G., Pines, D., and Meltzer, D. (eds.), *Complexity: Metaphors, Models, and Reality* (Reading, MA: Addison-Wesley, 1994).

Miller, J., and Page, S., *Complex Adaptive Systems* (Princeton, NJ: Princeton University Press, 2007).

Mitchell, M., *Complexity: A Guided Tour* (New York: Oxford University Press, 2011).

Part II: Getting Down to Cases

I drew material for this part from an unpublished article I wrote as part of an OECD study on global shocks, especially in the chapters on the Internet, pandemics, and deflation. The entire article can be found at:

Casti, J., "Four Faces of Tomorrow, "OECD International Futures Project on Future Global Shocks, OECD, Paris, January 2011 (www.oecd.org/data oecd/57/60/46890038.pdf).

62 Hempsell's ideas presenting the categories of events presented here are given in Hempsell, C. M., "The Potential for Space Intervention in Global Catastrophes," *Journal of the British Interplanetary Society, 57* (2004), 14–21.

A related article focusing mostly on the extinction-type of events is Bostrum, N., "Existential Risks," *Journal of Evolution and Technology, 9* (March 2002).

Digital Darkness: A Long-Term, Widespread Failure of the Internet

Unlike some of the other topics dealt with in Part II, the body of information available on Internet security truly boggles the mind. Moreover, at present the problem is undergoing a huge reexamination. So it would be pointless to present a long list of citations here, when most of them would have been superseded long before this book was even published. As a result, I have listed here only a couple of very general pointers, together with specific articles that formed the background to the major stories presented in the chapter itself. For a more updated account of the current state of play regarding changes in the Internet, I urge the reader to just search online under terms like "cybersecurity, cyberwarfare" and the like for more information than you could ever hope to digest.

Two very interesting and informative recent works about the Internet and its future are the volumes:

Zittrain, J., *The Future of the Internet* (New Haven, CT: Yale University Press, 2008).
Morozov, E., *The Net Delusion* (Cambridge, MA: Public Affairs, 2011).

The first book is a pretty balanced account of the pros and cons of the Internet as it stands today, outlining the fact that the "black hats" are gaining the upper hand, with spyware, viruses, and other malware beginning to choke off the huge communication and information-availability benefits of the Internet. The author then outlines a program for preserving the good, while closing down the bad. The book by Morozov argues the far more radical thesis that the entire notion of "Internet freedom" is an illusion. Technology has not made the world more democratic, but in fact has actually allowed authoritarian regimes to exert even more control of their citizens than ever before. Morozov asserts that we are all actually being pacified by the Internet rather than engaging with it. On balance, this volume argues a very provocative and thoughtful thesis, one that every Internet user should be aware of.

69 The story of **Dan Kaminsky's discovery** of the flaw in the DNS system is told in: Davis, J., "Secret Geek A-Team Hacks Back, Defends Worldwide Web," *Wired, 16,* no. 12 (November 24, 2008).

As an illustration of the vast amount of literature available on how the Internet might crash, the following article is representative (and note this was in 1997!): Garfinkel, S., "50 Ways to Crash the Internet," *Wired,* August 19, 1997.

I could list a dozen more recent articles all telling pretty much the same story. But the one above is the most fun, and very few of the fifty methods discussed have been effectively addressed even now, more than fifteen years after its publication.

84 The problem of **router scalability** is addressed at the following sites:

http://www.potaroo.net/ispcol/2009-03/bgp2008.html
http://blog.caida.org/best_available_data/2008/05/10/top-ten-things-lawyers-should-know-about-internet-research-8/

84 The story of **Schuchard's procedures** for taking down the Internet via botnets is chronicled in: Aron, J., "The Cyberweapon That Could Take Down the Internet" (http://www.newscientist.com/article/dn20113-the-cyberweapon-that-could-take-down-the-internet.html).

86 The **Stuxnet computer worm** was discovered by German cybersecurity expert Ralph Langner in 2010. His deep study of the peculiarities of this worm, together with the fact that the Stuxnet seemed to be targeted primarily at Iranian nuclear facilities, led Langner to make the bold claim that it is actually a nasty piece of software introduced by the US intelligence community that has gotten out of the control of its creators. For this story, see: Gjelten, T., "Security Expert: U.S. 'Leading Force' Behind Stuxnet," PBS, September 26, 2011 (http://www.npr.org/2011/09/26/140789306/security-expert-u-s-leading-force-behind-stuxnet).

88 An excellent overview of the entire **DOS problem** is given in: Mirkovic, J., et al. "Understanding Denial of Service," *InformIT,* August 12, 2005 (http://www.informit.com/articles/article.aspx?p=386163).

89 I'm afraid I cannot give a citation to **Noam Eppel's unbridled attack on Internet security**, since as I noted in the text, the site from which I obtained the paper has since been shut down and the paper itself seems to have turned into "vapor paper," so to speak.

Finally, the reader might want to have a look Dave Pollard's account of what life will be like after the Internet crashes. I didn't feature it in the text, but it's still an interesting view of how life will be when the Internet is no longer part of it. The citation is:

Pollard, D., "What Are You Going to Do When the Internet's Gone?" (www.howtosavetheworld.ca/2010/05/04/what-are-you-going-to-do-when-the-Internets-gone).

When Do We Eat: Breakdown of the Global Food-Supply System

94 A popular account of the **tree virus in the United Kingdom** is given in the following article: Middleton, C., "Red Alert in Britain's Forests as Black Death Sweeps In," *Daily Telegraph,* February 3, 2011.

More information about it can be found at the UK Forestry Commission website, www.forestry.gov.uk/pramorum.

96 John Christopher's eye-opening novel about the social effects from the **Chung-Li virus** destroying food crops worldwide was originally published in 1956. A recent reprint is: Christopher, J., *The Death of Grass* (London: Penguin, 2009).

97 The **"doomsday" seed vault** was well chronicled in the general press at the time of its opening in 2008. See, for example: Mellgren, D., "'Doomsday' Seed Vault Opens in Arctic." Associated Press release, 2008 (www.seed-vault.no).

98 The **death of bees** is a story well told in the book: Jacobsen, R., *Fruitless Fall* (New York: Bloomsbury, 2008).

Two of the many semipopular accounts of the two sides of the honeybee collapse story are portrayed in the following:

Aizen, M., and Harder, L., "The Truth About the Disappearing Honey-bees," *New Scientist*, October 26, 2009.

Armstrong, D., "Bee-Killing Disease May Be Combination Attack," *Bloomberg News*, October 7, 2010.

102 The story of **Ms. Galviso's odyssey** to find Thai jasmine rice for her family is told in: http://www.businessweek.com/bwdaily/dnflash/content/apr2008/db20080424_496359.htm.

The huge run-up in food prices worldwide has been featured in many articles and books. Among the more accessible for general readers are:

Brown, L., "The Great Food Crisis of 2011," *Foreign Policy*, January 10, 2011.

Sircus, M., "Food/Financial Crisis of 2011" (http://agriculture.imva.info/food-prices/foodfinancial-crisis-of-2011).

Wallop, H., "Global Food Prices Hit New Record High," CommonDreams.org, February 3, 2011 (www.commondreams.org/headline/2011/02/03-1).

Sen, A., "The Rich Get Hungrier," *New York Times*, May 28, 2008 (http://www.nytimes.com/2008/05/28/opinion/28sen.html).

103 The direct links between **rising food prices and political and social unrest** is another theme that's been well chronicled over the past few years. Two of the many sources contributing to the discussion in the text are:

Karon, T., "How Hunger Could Topple Regimes," *Time*, 2008 (www.time.com/time/world/article/0,8599,1730107,00.html).

Chang, G., "Global Food Wars," *New Asia*, February 21, 2011 (blogs.forbes.com/gordonchang/2011/02/21/global-food-wars.html).

104 A very enlightening overview of the effect **global warming** is having (and will continue to have) on food production is available in: Gillis, J., "A Warming Planet Struggles to Feed Itself," *New York Times*, June 4, 2011 (www.nytimes.com/2011/06/05/science/earth/05harvest.html).

The Day the Electronics Died: A Continent-Wide Electromagnetic Pulse Destroys All Electronics

Probably the most definitive source for material on the EMP as a threat to society is: *Report of the Commission to Assess the Threat to the United States from Electromagnetic Pulse (EMP) Attack, Volume 1. Executive Report* (Washington, DC: US Government Printing Office, 2004) (available at amazon.com).

Another good source is the volume: Gaffney, F., *War Footing: 10 Steps America Must Take to Prevail in the War for the Free World* (Annapolis, MD: US Naval Institute Press, 2005).

While there do not seem to be too many actual books (nonfiction, that is) focused on an EMP, there are volumes of items on the subject in professional and academic journals, as well as Internet postings addressing this threat. Here is a sampling that I found useful in preparing this chapter:

Kopp, C., "The Electromagnetic Bomb: A Weapon of Electrical Mass Destruction." In *Information Warfare—Cyberterrorism: Protecting Your Per-*

sonal Security in the Electronic Age, ed. W. Schwartau (New York: Thunder Mouth Press, 1996).

Spencer, J., "The Electromagnetic Pulse Commission Warns of an Old Threat with a New Face." Backgrounder #1784, The Heritage Foundation, Washington, DC, August 3, 2004.

"Electromagnetic Pulse Risks and Terrorism," United States Action Group (http://www.unitedstatesaction.com/emp-terror.htm).

Dunn, J. R., "The EMP Threat: Electromagnetic Pulse Warfare," April 21, 2006 (http://www.americanthinker.com/2006/04/the_emp_threat_electromagnetic.html).

116 The story of how both the United States and the USSR were thinking about an **EMP as an opening salvo** in an all-out nuclear exchange during the Cold War is told in: Burnham, D., "U.S. Fears One Bomb Could Cripple the Nation.," *New York Times,* June 28, 1983, p. 1.

117 As always, the science-fiction literature is way ahead of reality as there are several very entertaining and scary accounts available of **how life might look in the aftermath of an EMP attack.** One recent entry in this area is: Forstchen, W., *One Second After* (New York: TOR Books, 2009).

117 The **quotes from both the Chinese military commander and Iranian defense analyst N. Nezami** are given in the following volume by US defense analyst Frank J. Gaffney: Gaffney, F., *War Footing* (Annapolis, MD: Naval Institute Press, 2005).

A New World Disorder: The Collapse of Globalization

The phenomenon of globalization has been chronicled and rechronicled to a point where regardless of what view you hold on the matter, there is a book somewhere that will support your position. So let me just list a few of the publications I consulted in preparation of the discussion found in the text:

James, H., *The Creation and Destruction of Wealth* (Cambridge, MA: Harvard University Press, 2009).

Dumas, C., *Globalisation Fractures* (London: Profile Books, 2010).

Walljasper, J., "Is This the End of Globalization?," *Ode,* July 2004 (www.odemagazine.com).

124 Finally, here is the citation to **Saul's analysis** of the collapse of the entire process: Saul, J. R., *The Collapse of Globalism* (Victoria, Australia: Penguin, 2005).

126 The story of **Russia's future under Vladimir Putin** is characterized in much the same terms as outlined here in the following articles, which coincidentally appeared on the same day in the same venue:

Freeland, C., "Failure Seen in Putin's Latest Move," *New York Times,* September 29, 2011.

Charap, S., "In Russia, Turning Back the Clock," *New York Times,* September 29, 2011.

126 The story of **Ms. Volkova and the unhappy fate of the Russian liberals** is recounted in: Barry, E., "For Russia's Liberals, Flickers of Hope Vanish," *New York Times,* September 25, 2011.

129 The story of the **collapse of the European Union** as being driven by the current downward movement of the collective social mood in Europe is given in: Casti, J., *Mood Matters: From Rising Skirt Lengths to the Collapse of World Powers* (New York: Copernicus, 2010).

131 The scenarios painted in the text for a **collapse of the euro** are outlined in the following article: Phillips, J., and P. Spina., "What Will Happen to Currencies If the Euro Collapses?," *Gold Forecaster Bulletin,* April 30, 2010.

132 **Project Proteus** is described for a public audience in the book: Loescher, M. S., Schroeder, C., and Thomas, C. W., compiled by Krause, P., *Proteus: Insights from 2020* (Washington, DC: Copernicus Institute Press, 2000).

135 The **Global Business Network scenarios** for the end of America are given in: Levin, J., "How Is America Going to End?," *Slate,* August 3, 2009 (www.slate.com/id/2223962/).

138 **Niall Ferguson's rousing endorsement** of complex systems theory in the cause of historical analysis is found in: Ferguson, N., "Empires on the Edge of Chaos," *Foreign Affairs,* March/April 2010 (www.informationclearinghouse.info/article24874.htm).

140 The **Bloomberg piece on bunkers for oligarchs** is: Pronina, L., "Apocalypse Angst Adds to Terrorist Threat as Rich Russians Acquire Bunkers," *Bloomberg News,* May 12, 2011.

142 *Fortune* **magazine's list of the "new normals"** is available in full form as: Tseng, N.-H., "Five 'New Normals' That Really Will Stick," *Money Magazine,* August 23, 2010 (http://money.cnn.com/2010/08/20/news/economy/New_normal_economy.fortune/index.htm).

Fear of Physicists: Destruction of the Earth Through the Creation of Exotic Particles
An absolute must-read account of the possibility of accelerator disasters wiping out the world is given by Nobel Prize–winning physicist Frank Wilczek in: Wilczek, F., "Big Troubles, Imagined and Real." In *Global Catastrophic Risks,* ed. N. Bostrum and M. Cirkovic (Oxford: Oxford University Press, 2008), pp. 346–362.

Two more stimulating discussions of the same theme are given within the context of a broader array of X-events in the volumes:

Rees, M. *Our Final Century* (London: Arrow Books, 2003).
Posner, R. *Catastrophe: Risk and Response* (Oxford: Oxford University Press, 2004).

A somewhat more detailed and academic discussion of the history of physics experiments threatening the world and ways to assess the risk of such happenings is the article: Kent, A., "A Critical Look at Risk Assessments for Global Catastrophes," *Risk Analysis, 24,* no. 1 (2004), 157–168.

147 A fascinating account of the **origin of Higgs's ideas** and the enormous scientific and political obstacles that had to be overcome to get the Large Hadron Collider built to search for the "God particle" is found in the book: Sample, I., *Massive: The Hunt for the God Particle* (London: Virgin Books, 2011).

151 An interesting journalistic account of the supposed **strangelets that ran through the earth in 1993** is found in the article: Matthews, R., "Earth Punctuated by Tiny Cosmic Missiles," *London Daily Telegraph,* November 5, 2002.

153 A fascinating sociological study of the clash between science and the public inter-

est in regard to the **Brookhaven RHIC brouhaha** is given in the article: Crease, R., "Case of the Deadly Strangelets," *Physics World,* July 2000, 19–20.

155 A short, but complete, history of the **development of the Large Hadron Collider** is found in the BBC article: "Building the 'Big Bang' Machine," *BBC News,* April 9, 2008 (http://news.bbc.co.uk/go/pr/fr/1/hi/sci/tech/7595855.stm).

It's interesting to see what the world's top physicists think will come bursting forth from the LHC. Just before the machine was officially turned on, *Newsweek* magazine did just such a survey, asking people like Stephen Hawking, Brian Greene, and Steven Weinberg, among others, just what they imagined would be discovered. Their replies can be found in: "Forecasting the Fate of Mysteries," *Newsweek,* September 6, 2008.

Blown Away: Destabilization of the Nuclear Landscape

163 In the 1950s thru the late 1960s, the RAND Corporation was about as exciting an intellectual environment as existed anywhere. Not only were people like **Herman Kahn** thinking about the nuclear matters discussed here, mathematicians were developing new tools like game theory, linear and dynamic programming, and network flow analysis to solve the optimization problems posed by these very practical military matters. In addition, economists and other social scientists were working on what has become known as cost-benefit analysis, Delphi methods for glimpsing the future, and a whole host of other themes that were unheard of at the time but are now commonplace. An interesting account of these days is found in the volume: Smith, Bruce R., *The RAND Corporation: Case Study of a Nonprofit Advisory Corporation* (Cambridge, MA: Harvard University Press, 1966).

Herman Kahn's controversial book on nuclear Armageddon is: Kahn, H., *On Thermonuclear War* (Princeton, NJ: Princeton University Press, 1960).

An entertaining and informative biography of Kahn's work, including an extended account of his later gurulike status as a futurologist, is: Ghamari-Tabrizi, S., *The Worlds of Herman Kahn* (Cambridge, MA: Harvard University Press, 2005).

164 As good a place as you'll find for learning the basic ideas underlying the **MAD strategy of nuclear deterrence** is the following volume by Nobel Prize winner Thomas C. Schelling: Schelling, T., *Strategy of Conflict* (Cambridge, MA: Harvard University Press, 1960).

166 Along with the *Ambio* **article by Birks and Crutzen** cited in the text, the spark that set off the heated nuclear winter debate is the book: Ehrlich, P., et al., *The Cold and the Dark: The World After Nuclear War* (New York: Norton, 1984).

Another volume of the same period that's worth a look is: Greene, O., Percival, I., and Ridge, I., *Nuclear Winter* (Cambridge, UK: Polity Press, 1985).

The TTAPS article by Carl Sagan and his colleagues that established the basis for the science of nuclear winter is: Turco, R., et al., "Global Atmospheric Consequences of Nuclear War," *Science, 222* (1983), 1283ff.

Many more accounts of the overall phenomenon of nuclear winter, updated scenarios, and forecasts are readily available via a web search. They are not listed here as the overall conclusion is unchanged in broad outline, if not detail, from the original work done in the 1980s.

171 More details on the **hypothetical New York City terrorist nuclear attack** are found at the website www.atomicarchive.com, while the website www.carol moore.net is a treasure trove of material on various nuclear scenarios, including the Israel-Iran scenario reported here.

For more details on paradoxes of nuclear safety, the following volume is difficult to beat: Leslie, J., *The End of the World* (London: Routledge, 1996).

Running on Empty: Drying Up of World Oil Supplies

There are almost as many books, articles, videos, and other sorts of materials available about the peak oil problem and the impending "end of oil" as there are people with an opinion on the matter. So the following list of sources is nothing but a small tip of a huge iceberg. But it's a tip that I found useful in assembling the story presented here, and each item contains numerous follow-up references for the interested reader to consult. To begin, let me note that some of the material in this chapter is adapted from my earlier book, *Mood Matters*, cited earlier.

An excellent, if slightly hyper, exposition of what we can expect from the end of oil and when is the volume: Kunstler, J., *The Long Emergency* (New York: Atlantic Monthly Press, 2005). This well-researched volume starts with peak oil and goes on to detail the numerous infrastructure collapses and lifestyle changes that will follow from it. I highly recommend this book to anyone looking for a one-stop-shopping account of the end of the petroleum age and how humanity can survive it.

Here are a few other volumes that mine the same territory:

Goodstein, D., *Out of Gas* (New York: Norton, 2004).
Middleton, P., *The End of Oil* (London: Constable and Robinson, 2007).
Strahan, D., *The Last Oil Shock* (London: John Murray, 2007).

An overview of the end of not only oil but numerous other commodities is: Heinberg, R., *Peak Everything* (Forest Row, UK: Clairview Books, 2007).

A very good look at the entire peak oil scenario in question-and-answer format is: "Life After the Oil Crash," www.salagram.net/oil-in-crisis.htm.

The following volume provides the background of Hubbert's famed 1956 forecast of the US peak oil moment, as well as a look at the current global situation: Deffeyes, K., *Hubbert's Peak* (Princeton, NJ: Princeton University Press, 2001).

I'm Sick of It: A Global Pandemic

The Plague was one of the Camus's most influential novels and almost surely figured heavily in his being awarded the Nobel Prize for literature in 1957. It has been reprinted many times so I won't bother listing the book here. But two somewhat more recent fictional accounts of what could happen if unknown pathogens get loose are worth a look. They are:

Preston, R., *The Cobra Event* (New York: Ballantine Books, 1997).
Ouellette, P., *The Third Pandemic* (New York: Pocket Books, 1997).

The Internet is full of accounts of pandemics and plagues. One that I found extremely useful is from a course at Hartford University: History of Epidemics and Plagues (http://uhavax.hartford.edu/bugl/histepi.htm).

An outstanding general reference giving the "big picture" on epidemics is available at the website of the Wellcome Trust in the United Kingdom: www.wellcome.ac.uk/bigpicture/epidemics. Material from this document served as the basis for several of the stories recounted in this chapter.

198 **Nathan Wolfe and his colleagues** have argued that major diseases of food-producing humans are of relatively recent origin, having originated only in the

Díganos acerca de su visita a Walmart hoy y usted podría ganar una de las 5 tarjetas de regalo de Walmart de $1000 o una de las 750 tarjetas de regalo de Walmart de $100.

http://www.survey.walmart.com

No purchase necessary. Must be 18 or older and a legal resident of the 50 US, DC, or PR to enter. To enter without purchase and for official rules, visit www.entry.survey.walmart.com.

Sweepstakes period ends on the date outlined in the official rules. Survey must be taken within ONE week of today. Void where prohibited.

THANK YOU

How was your experience?

Tell us about your visit today and you could win 1 of 5 $1000 Walmart gift cards or 1 of 750 $100 Walmart gift cards.

Díganos acerca de su visita a Walmart hoy y usted podría ganar una de las 5 tarjetas de regalo de Walmart de $1000 o una de las 750 tarjetas de regalo de Walmart de $100.

http://www.survey.walmart.com

No purchase necessary. Must be 18 or older and a legal resident of the 50 US, DC, or PR to enter. To enter without purchase and for official rules, visit www.entry.survey.walmart.com.

Sweepstakes period ends on the date outlined in the official rules. Survey must be taken within ONE week of today. Void where prohibited.

See back of receipt for your chance
to win $1000

Walmart ><

Save money. Live better.

(904) 797 - 3309
MANAGER GARY ANDERSON
2355 US HIGHWAY 1 S
ST AUGUSTINE FL 32086
STW 00579 OP# 000114 TE# 52 TR# 01003
HP 63 BLACK   088929626746   17.97 X
                    SUBTOTAL   17.97
       TAX 1   6.500 %   1.17
                    TOTAL   19.14
         SHOPPING CARD TEND   19.14
                 CHANGE DUE   0.00

SHOP.CARD REDEMPTION   19.14
ACCOUNT 613374701048
APPR. CODE = 087161
REF #0866875
Beg Bal   Tran Amt   End Bal
191.25   19.14   172.11
04/16/17   11:04:42

                 # ITEMS SOLD 1
     TC# 5914 1000 1172 6961 3027

     04/16/17   11:04:44
Store receipts on your phone. Walmart P
ay.

last eleven thousand years. Their argument is presented in: Wolfe, N., Dunavan, C., and Diamond, J., "Origins of Major Human Infectious Diseases," *Nature, 447* (May 17, 2007), 279–283.

199 The **Ebola fever story** is recounted in the best-seller: Preston, R., *The Hot Zone* (New York: Random House, 1994).

200 A complete account of **Gladwell's three laws of epidemics** is found in his immensely thought-provoking and entertaining popular book: Gladwell, M., *The Tipping Point* (London: Little, Brown, 2000).

204 **Typhoid Mary's sad story** is available at dozens of websites. The Wikipedia entry under "Mary Mallon" is a good place to start.

207 A detailed discussion of the threat posed by **avian flu** is provided by Mike Davis in his book: Davis, M., *The Monster at Our Door* (New York: The New Press, 2007).

209 The work described in the text using the **World of Warcraft** as a virtual world for studying the spread of epidemics is published as: Lofgren, E., and N. Fefferman, "The Untapped Potential of Virtual Game Worlds to Shed Light on Real World Epidemics," *The Lancet. Infectious Diseases, 7,* no. 9 (September 2007), 625–629.

Another Internet world being used for the same type of work is the game "Where's George?" in which players keep track of the movement of dollar bills as they travel about the world. An account of this work is provided in the science blog "Web Game Provides Breakthrough in Predicting Spread of Epidemics," www.scienceblog.com/cms.

Dark and Dry: Failure of the Electric Power Grid and the Disappearance of Clean Water Supply

Of the many popular and semipopular volumes on the electrical power grid, two that I found especially enlightening and useful are:

> Makansi, J., *Lights Out* (New York: Wiley, 2007).
> Schewe, P., *The Grid* (Washington, DC: Joseph Henry Press, 2007).

Both of these books give a vivid account of the history of the development of the power grid, the largest industrial investment in history, and arguably the greatest engineering achievement, as well. They each portray not only the fascinating history of the grid, but also its many vulnerabilities and the consequences for daily life of ignoring them.

A good discussion of how the electric power grid must be changed to meet societal demands over the next decades is given in: Gellings, C., and K. Yeager, "Transforming the Electric Infrastructure," *Physics Today, 57* (December 2004), 45.

There are numerous extensive accounts on the Internet of the various power blackouts mentioned in the text, so I won't recount them here. But due to its unfolding character and threat to the economy of an entire country, it's worth citing a couple of references to the situation in South Africa: Mnyanda, L., and Theunissen, G., "Rand Sinks as South African Electricity Grid Fails," *Bloomberg.com,* February 11, 2008.

The reader should also see articles in the *International Herald Tribune* published on January 30–31, 2008.

Two very useful references on the water shortage crisis are the volumes: Pearce, F., *When the Rivers Run Dry* (London: Eden Project Books, 2006); and Clarke, R., and J. King, *The Atlas of Water* (London: Earthscan Books, 2004).

An important question is how to see the upcoming water shortage in a rational manner, so as to provide adequate clean water for future needs. This question is addressed head-on in the article: Smil, V., "Water News: Bad, Good and Virtual," *American Scientist*, September–October 2008, 399–407.

For a set of eye-opening graphics displaying the water shortage situation, see those in the posting "Drought" (www.solcomehouse.com/drought.htm).

227 The story of the **UK water shortage** caused, paradoxically, by the massive flooding in 2007, is recounted in: Elliott, V., "Looting, Panic Buying—and a Water Shortage," *Times Online,* July 23, 2007 (http://www.timesonline.co.uk/tol/news/uk/article2120922.ece).

230 **Professor Tony Allan** received the Stockholm Water Prize in 2008, a prestigious award from the Stockholm Water Foundation, for outstanding water activities and research. An account of the virtual water concept is given in the announcement of this award from the SIWI at www.siwi.org/sa/node.asp?node=25.

Technology Run Amok: Intelligent Robots Overthrow Humanity

233 For the **Gordon Moore** quote that opens the chapter, see the Wikipedia entry under "Moore's Law": http://en.wikipedia.org/wiki/Moore%27s_law.

235 The definitive work outlining all aspects of **the Singularity problem** is the volume: Kurzweil, R., *The Singularity Is Near* (New York: Penguin, 2005).

A slightly earlier book by the science-fiction writer Damien Broderick, who terms the Singularity the "Spike," mining much the same territory but with a more social perspective, is: Broderick, D., *The Spike* (New York: TOR Books, 2001).

The starting point for the entire idea of a technological singularity is the following 1993 paper by mathematician and sci-fi writer, Vernor Vinge: Vinge, V., "The Coming Technological Singularity: How to Survive in the Post-Human Era." Paper presented at the VISION-21 Symposium, NASA Lewis Research Center, March 30–31, 1993. (See also a slightly revised version of the paper in the Winter 1993 issue of *Whole Earth Review*.)

238 The guru of the **nanotech movement** is physicist K. Eric Drexler, who has presented his vision of the future in the following works:

> Drexler, K. E., *Engines of Creation: The Coming Era of Nanotechnology* (New York: Doubleday, 1986).
> Drexler, K. E., *Nanosystems: Molecular Machinery, Manufacturing and Computation* New York: Wiley, 1992).

238 A fascinating account of why the **"gray goo" scenario** for the end of the world is extremely unlikely is given in the paper: Freitas, R., "Some Limits to Global Ecophagy by Biovorous Nanoreplicators, with Public Policy Recommendations," *The Foresight Institute,* 1991 (www.foresight.org/nano/Ecophagy.htm).

238 An interesting account of **how AI impacts the global risk situation** is presented in the chapter: Yudkovsky, E., "Artificial Intelligence as a Positive and Negative Factor in Global Risk." In *Global Catastrophic Risk,* ed. N. Bostrom and M. Cirkovic (Oxford: Oxford University Press, 2008), 308–346.

244 **Asimov's three laws of robotics** are presented, along with the Fourth Law discussed in the text and a detailed discussion of the entire intelligent robot question in the article: Branwyn, G., "Robot's Rules of Order" (http://www.informit.com/articles/article.aspx?p=101738).

Another very detailed discussion of this theme is given by Roger Clarke at the site http://www.roger/- clarke.com/SOS/Asimov.html.

241 Bill Joy's cri de coeur outlining the dangers of **the GNR problem** is found in: Joy, W., "Why the Future Doesn't Need Us," *Wired,* April 2000.

The Great Unwinding: Global Deflation and the Collapse of World Financial Markets
Bookshelves sag under the weight of volumes of various sorts all purporting to describe the Great Recession of 2007-08, along with how the world's economic fate is likely to unfold in the coming decades. Oddly enough, it's difficult to find even one of these learned and/or popular tomes that even mentions deflation as a candidate for the economic profile of the near-term future. The arguments made here seem to be essentially ignored by the pundits and the economic cognoscenti, which, given their record in forecasting anything like what actually takes place, seems as strong a reason as any to look long and hard at a deflationary scenario. The one volume I have at hand that actually does address this very real possibility is by Nouriel Roubini and Stephen Mihm. Given that Roubini has achieved near-mythical status in some quarters for his anticipation of the Great Recession, his serious treatment of deflation as a viable candidate for the upcoming global economy should in my view be taken very seriously indeed. The full reference is: Roubini, N., and S. Mihm., *Crisis Economics* (New York: Penguin, 2010).

251 **Damon Vickers's scenario** for the big crash is found in his very entertaining and scarifying book: Vickers, D., *The Day After the Dollar Crashes* (New York: Wiley, 2011).

254 **Seabright's article** on the structures put in place to supposedly insulate the economy from another 1930s-style crash is available in: Seabright, P., "The Imaginot Line," *Foreign Policy,* January–February 2011.

255 The quote attributed to **Robert Lucas** about economists having solved the problem of the Great Depression is given in the absolutely first-rate article by Paul Krugman addressing the issue of how academic economics has gone so far off the track: Krugman, P., "How Did Economists Get It So Wrong?," *New York Times Magazine,* September 2, 2009 (http://www.nytimes.com/2009/09/06/magazine/06Economic-t.html?page-wanted=all).

Some of the material mentioned in the opening salvos in the text regarding the **onset of the financial crisis** include:

> Samuelson, R., "Rethinking the Great Recession," *Wilson Quarterly,* Winter 2011, 16–24.
> Krugman, P., "A Crisis of Faith," *New York Times,* February 15, 2008.
> Thompson, D., and D. Indiviglio, "5 Doomsday Scenarios for the U.S. Economy," *The Atlantic,* September 2, 2010.

261 The quote from **Steve Hochberg** about deflation appeared in *Elliott Wave Short-Term Financial Forecast,* Elliott Wave International, Gainesville, GA, September 8, 2011.

262 Two excellent one-stop-shopping **explanations of deflation** for the uninitiated are:

> Hendrickson, M., "Demystifying Deflation," *American Thinker,* October 12, 2010 (www.american-thinker.com/archived-articles/2010/10/demystifying_deflation.html).

A Visual Guide to Deflation (www.mint.com/blog/wp-content/uploads/2009/04/visual-Guidetodeflation).

265 The **social mood of a society** is a major factor in biasing the sorts of social events one can expect to see. This point is elaborated in great detail in the book: Casti, J., *Mood Matters: From Rising Skirt Lengths to the Collapse of World Powers* (New York: Copernicus, 2010).

266 The sad **Japanese experience** in living in a deflationary environment is chronicled in the following articles:

Fackler, M., "Japan Goes from Dynamic to Disheartened," *New York Times,* October 16, 2010.
Suess, F., "2010 And Beyond—Deflation, Japanese Style," *Daily Bell,* January 16, 2010 (www.thedailybell.com).

267 The quote from **Richard Koo on Japanese deflation** is taken from his fantastic book describing the entire process: Koo, R., *The Holy Grail of Macroeconomics: Lessons from Japans Great Recession* (New York: Wiley, 2009).

Part III: X-Events Redux

276 The work by Stephen Carpenter and his group from the University of Wisconsin on **identifying early-warning signals of the collapse of the lake ecosystem** is described in:

Keim, B., "Scientists Seek Warning Signals for Catastrophic Tipping Points," *New York Times,* September 2, 2009.
Sterling, T., "Scientists Detect Early Warning of Ecosystem Collapse in Wisconsin," *The Cutting Edge,* May 2, 2011 (http://www.thecuttingedge news.com/index.php?article=51948&pageid=28&pagename=Sci-Tech).

The definitive citation for the full story is: Carpenter, S.R., et al., "Early Warnings of Regime Shifts: A Whole-Ecosystem Experiment," *Science,* April 28, 2011.
Other recent works covering an even broader scope of issues surrounding early-warning signals, including the area of climate change are:

Dakos, V., *Expecting the Unexpected,* Doctoral thesis, University of Wageningen, Wageningen, The Netherlands, 2011.
Dakos, V., et al., "Slowing Down as an Early Warning Signal for Abrupt Climate Change," *Proceedings of the National Academy of Sciences, 105* (September 23, 2008), 14308–14312.

278 A layman's account of some of the **dynamical-systems-based approaches** for anticipation of X-events is presented in: Fisher, L., *Crashes, Crises, and Calamities* (New York: Basic Books, 2011).

282 **Computer-based tools for analyzing** "What if . . . ?" types of questions looking for early-warning signals of big things to come are treated in:

Casti, J., *Would-Be Worlds* (New York: Wiley, 1997).

Epstein, J., *Generative Social Science* (Princeton, NJ: Princeton University Press, 2006).

Ehrentreich, N., *Agent-Based Modeling* (Heidelberg: Springer, 2008).

Gilbert, N. *Agent-Based Models* (Los Angeles: Sage Publications, 2008).

285 The **China scenarios** and results of the World Trade Network Simulation model's answers to the questions the scenarios raise are discussed in the following volume, which is the final report for the Game Changers project carried out by the author and colleagues for a consortium of Finnish and Scottish governmental agencies and private concerns in 2010–11. The citation is: Casti, J., et al., *Extreme Events* (Helsinki: Taloustieto Oy, 2011).

287 The **fate of Air France 447** given in the text follows the account presented in: Schlangenstein, M., and M. Credeur, "Air France Crew May Have Faced Baffling Data," *Bloomberg News,* May 28, 2011.

289 The **Bookstaber account of the complexity gap** between the SEC and the financial markets is found in his very accessible and entertaining book: Bookstaber, R., *A Demon of Our Own Design* (New York: Wiley, 2007).

292 The idea of **"willful blindness"** as a focal topic for why humans have such a predilection for taking actions that are manifestly against their own best interests is very timely today. Margeret Heffernan's book of the same title shows this phenomenon in every aspect of life, ranging from investment in Ponzi schemes to the war in Iraq: Heffernan, M., *Willful Blindness* (New York: Doubleday, 2011).

293 **Stiglitz's argument** for the growing complexity gap between the rich and the poor in American life is found in: Stiglitz, J., "Of the 1%, By the 1%, For the 1%," *Vanity Fair,* May 2011.

296 **Disaster myths as an indication** for how the public at large will react to extreme events is explored in the article: Schoch-Spana, M., "Public Responses to Extreme Events—Top 5 Disaster Myths," *Resources for the Future,* October 5, 2005 (http://www.rff.org/rff/Events/upload/20180_1.pdf).

297 The idea that **it takes an X-event to shock a system** out of a complexity gap is explored implicitly in the following article about Japan and the March 2011 earthquake: Pesek, W., "Roubini Earthquake Gloom Meet 'Shock Doctrine,'" *Bloomberg News,* March 13, 2011 (http://www.bloomberg.com/news/2011-03-13/roubini-earthquake-gloom-meets-shock-doctrine-william-pesek.html).

INDEX

Abu Ghraib, 292–93
accidental nuclear strikes, 174
accurate missiles, 178
Ackerman, Thomas, 168–70
adaptation, 300–301
adaptive agents, in agent-based simulations, 283–84
affluence, and food supply, 105, 106
agent-based simulation, 280–87
aggressive nuclear strikes, 174
AIDS/HIV, 197, 198, 205, 206
Air France Flight 447, 287–89
air traffic control systems, 25–26
Allan, Tony, 230, 314n
al-Qaeda, 115, 117, 188–89
alternating current (AC), 222–23
Amazon.com, 50–51
Ambrose, Stanley, 21, 304n
analog vs. digital, 241–42
anatomy of an X-event, 273–75
Andromeda Strain, The (Crichton), 29–31
anticipating X-events, 24, 40–41, 273–74; mathematical tools for, 278–80, 316n
Antonine Plague, 197
Arab Spring, 9–13, 27, 53; Internet shutdown, 74–75, 86. *See also specific countries*
Arbesman, Samuel, 6, 14
arms reduction, 178
ARPAnet, 71
artificial intelligence, 239–46, 314n
Ashby, W. Ross, 56, 304n
Asimov, Isaac, 244–45
Athanassoulas, Alexander, 56

avian flu, 23, 203, 207–8, 313n

Bankers Trust, 290
Barabasi, Albert-Làszló, 202
Bardhan, Pranab, 124–25
baseball hitting streaks, 6–7
Becket, Thomas, 257–58
bee colony collapse disorder, 23, 98–102, 307–8n
bell curve, 34–37, 35n
Ben Ali, Zine El Abidine, 53–54
Bengal famine of 1943, 105
Berg, Bryan, 1–2, 303–4n
bifurcation theory, 277
Big Bang, 148, 151–55
bioterrorism, 211. *See also* pandemics
bird flu. *See* avian flu
Birks, John, 166–67, 168
Black, Fischer, 37
Black Death, 62, 198
black holes, 81–82, 149–52, 157–59
Black-Scholes formula, 37
Black Swan, The (Taleb), 37–40
black swans, 37–40
Bookstaber, Richard, 289–91
Bouillard, Alain, 288
Brazil, 72–73, 285, 286
Brock, William "Buz," 277
Buchanan, Pat, 55
budget deficit, US, 260–61, 268–69
Buffett, Warren, 261–62
bunkers, 140–41
"bureaucratic creep," 129
Burt's Bees, 100

Bush, George W., 55, 192
butterfly effect, 54–55, 72, 126–27, 174
Byzantine Empire, 45–46

cable fragility, 83
Cambrian explosion, 138
Campbell, Colin, 184–85
Camus, Albert, 195–97, 312n
Capone, Al, 126
Carpenter, Stephen, 276–78, 316n
Carr, Nicholas, 292
Carroll, Lewis, 49
Carson, Rachel, 100
"Cassandra effect," 292–93
Cat's Cradle (Vonnegut), 31, 32, 145, 156
central banking, 255–56, 257
CERN (European Nuclear Research
    Center), 146–51, 155–59
Chako Paul, 72
Cheney, Dick, 192
Chicken Little, 19
China: agent-based simulation, 285–87,
    317n; dust bowls, 104; financial system,
    52–53, 124; globalization, 49–50;
    Internet outage, 73–74; Japanese tragedy
    and, 297; nuclear landscape, 173; public
    outbreaks, 206; water shortages, 104, 106
Christopher, John, 96–97, 307n
Chung-Li virus, 96–97, 307n
cigarette smoking, 203–4
Clarke, Roger, 244–45
climate change. *See* global warming
Clinger, Gilbert, 110
Clinton, Bill, 188
Club of Rome, 193–94
Cobb, Ty, 7
coffeemakers, 41–42
Cold War, 161–66
Collapse of Complex Societies, The (Tainter),
    44, 45, 49
Collapse of Globalism, The (Saul), 124–25,
    309n
collision, systems in, 9–15
colony collapse disorder (CCD), 98–102,
    307–8n
"complexity creep," 51
complexity gaps, 14–15; characterizing and
    measuring, 36–37, 287–88
complexity overload, 24, 42, 44–46, 62,
    162, 216, 304–5n

complexity principles, 47–58
complexity trap, 1–4
Compton effect, 112
computer code, 42–43
consumer spending, 259–60
Conway, Kim, 220
cosmic rays, 159
"creative destruction," 269
credit default swaps (CDS), 256–57
Crichton, Michael, 29–31
critical slowing down, 279–80
cropland, 104, 106
Crutzen, Paul, 166–67, 168
currencies (currency markets), 106, 124,
    251–52. *See also* dollars, US; euro cur-
    rency
Darby, Joe, 292–93
dark matter, 150, 152
Day After the Dollar Crashes, The (Vickers),
    251–52, 253
Death of Grass, The (Christopher), 96–97,
    307n
deduction, 53
Defense Department, US, 70–71, 135,
    162
deflation, 143, 262–69, 315–16n
deflationary spiral, 262–68
deliberate Internet attacks, 85–89
Demon of Our Own Design, A (Booksta-
    ber), 289–91
deposit insurance, 254–55
deregulation, 224, 257
deteriorating transmission grid, 224
Diamond, Jared, 198–99, 305n
Dickenson, Ray, 88
DiMaggio, Joe, 6–7, 14
dinosaurs, 21–22
direct current (DC), 222–23
disaster myths, 295–97, 317n
Dixon, Paul, 154
Doctor Strangelove (movie), 163
dollars, US, 106, 124, 131, 141, 251–52,
    253, 260, 262, 268
Domain Name Server (DNS), 68–71, 75,
    87–89
doomsday seed vault, 97–98, 102
DOS attacks, 69–71, 75, 84–85, 87–89
droughts, 228
drugs, 205, 207
Duffy, Hugh, 7

Dunavan, Claire, 198–99
dust bowl, 104
Dutch elm disease, 94

early-warning principles, 278–80
East-Midwest Blackout of 2003, 215–16
Ebola fever, 199–200, 313n
"edge of chaos," 51
Edison, Thomas, 222–23
Egypt, 9, 54, 129; Internet shutdown, 11, 74–75, 83, 86
Einstein, Albert, 157
Eldredge, Nils, 137–38
electric power grid, 219, 221–23, 313n; history of, 222–23; vulnerabilities in system, 223–26
electric power grid failures, 212–26; adding it all up, 231–32; East-Midwest Blackout of 2003, 215–16; Great Northeast Blackout of 1965, 213–15, 225; New York City blackout of 1977, 212–13, 215
electromagnetic pulse (EMP), 109–21, 308–9n; adding it all up, 120–21; appropriate response to attack, 117–18; possible terrorist scenario, 115–16; preventing attack, 119–21
Eliot, T. S., 257–58, 295
e-mail, 70, 71–72, 78
emergent behaviors or traits, 48–49, 282
EMP Threat Commission, 118
endemics, 197
Enemy Within scenario, 133–34
engineers, 225
epidemics, 197
Eppel, Noam, 89–90, 307n
Estonia, Internet attack, 73, 88
ethanol, 105–6, 107
euro currency, 129–32, 253, 260–61
European Central Bank (ECB), 129–30, 131, 258
European Union (EU), 127–32, 260–61, 285, 310n
evolution, 137–38
exotic particles, 144–60, 310–11n
extensively drug-resistant TB (XDR-TB), 205, 207
extinction-level events, 62–63
extra dimensions, 157–58
extrasociety dilemma, 204–5

extreme events, 8–9, 13, 34–35, 62–63, 304n
extreme impact of X-events, 5, 38, 39
extreme weather, 33, 104, 107

Facebook, 11–12, 82, 225
Fackler, Martin, 267
Faraday cages, 120
fat-tailed distributions, 35–37, 35n
Federal Aviation Administration (FAA), 25–26
Federal Deposit Insurance Corporation (FDIC), 254–55
Federal Reserve, 38, 106, 124, 255–56, 257, 263–64
Fefferman, Nina, 208–10
Ferguson, Niall, 137–40, 310n
Fermi, Enrico, 146
financial crisis of 2007–2008, 37–38, 43, 106, 123–24, 255–61, 289–90, 295, 315n. See also global financial crisis
financial markets: emergent aspect, 48; history of, 254–57; Internet dependence, 78–79; role of trust, 254, 258–59
financial regulations, 255–56, 257, 291
Financial Stability Oversight Board, 289
first law of robotics, 244–45, 314n
Florida, presidential election of 2000, 218
flu (influenza), 30, 202–3. See also Spanish flu outbreak of 1918
fluctuations, increasing rate of, 278
flux compression generator (FCG), 112–13, 116, 121
food industrialization, 93, 97
food prices, 92, 102–6, 308n
food-supply system. See global food-supply system
food web, Peter Lake experiment, 276–77
football betting, 281–82
Forbidden Planet (movie), 236, 243–44
forecasting models, 36–37, 280–87
"forecasting" of X-events, 39–41, 54
foreclosures, 259
Frederick, Ivan, II, 292
Friedman, Thomas, 122, 124
Fukushima Daiichi nuclear disaster, 12–13, 256

Galviso, Mary Ann, 102–3
game theory, 163–64, 311n

genetic engineering, 237
genetic modification of crop seed, 101
Georgia (country), Internet attack, 88
Germany, 130–31, 218–19
Giannini, Ruggero, 181
Gibbons, Ann, 21
Gibson, William, 96
Gladwell, Malcolm, 200–201, 313n
Glass-Steagall Act, 257
Global Business Network (GBN), 135–37
global catastrophes, overview, 62–63
global disasters, overview, 62–63
global financial crisis, 123–24, 251–69,
    315–16n; adding it all up, 269; black
    swans and, 37–38; deflation and,
    262–69; economic defenses against,
    254–56; potential scenarios, 251–53,
    259–60; role of trust, 254, 258–59
global food-supply system, 92–108, 307n;
    adding it all up, 107–8; death of bees,
    98–102; decline in supply, 102–6; plant
    viruses, 94–98; seed banks, 97–98,
    102; "solution" to food crisis, 106–8
globalization, 46, 122–43, 309n; adding it
    all up, 140–43; caterpillar vs. butterfly,
    125–27; decline and fall of European
    Union, 127–32; empire transitions,
    137–40; GBN scenarios, 135–37; Pro-
    teus scenarios, 132–35, 136–37; Red
    Queen hypothesis, 49–50
global pandemics. See pandemics
global warming, 274–75; food supply and,
    104, 107, 308n; Internet data centers
    and, 82; nuclear exchange and, 169–70;
    water supply and, 227
Glos, Michael, 219
GNR problem, 237–39, 247
Gödel, Kurt, 53, 54
gold, 131, 141, 253
Goldilocks Principle, 51–53
Google, 11–12, 82
Gorbachev, Mikhail, 125
Gore, Al, 55, 192
Gould, Stephen J., 137–38
gray goo scenario, 32, 63, 238, 314n
Great Depression, 142, 258, 264, 315n
Great Northeast Blackout of 1965,
    213–15, 225
Great Recession of 2007–2008, 13–14,
    141–43, 255–61, 315n

Greece, 27, 56, 128, 253, 265
greenhouse gases, 226
Greenspan, Alan, 124, 258
Gumede, William Mervin, 217

hackers, 81, 85–89
Haiti earthquake of 2010, 28
Hanson, Jason, 141
hardware failures, 81–85
Hawking, Stephen, 152, 154, 311n
Hawtin, Geoff, 98
Healing, Richard, 288
health care system, and Internet depen-
    dence, 79–80
"hedge," 290
hedge funds, 257, 258
hedger's dilemma, 294–97
Heffernan, Margaret, 291–93, 317n
Hempsell, C. M., 62–63, 306n
Hengchun earthquake of 2006, 83
H5N1 virus, 203, 207–8
Higgs, Peter, 147–48, 310n
Higgs boson, 147–48, 155
high-amplitude fluctuations, 278–79
Hiroshima, 20, 170
historical processes, theories of, 137–40
Hochberg, Steven, 261–62
Homeland Security Department, US, 219
home ownership, 142
Homer-Dixon, Thomas, 45, 303n
housing market, 259
Hubbert, M. King, 182–83, 312n
human brain vs. computer, 240–43
human-caused X-events, overview, 22–32
hurricanes, 39, 296
Hurricane Houston, 188–89
Hurricane Irene, 19
Hurricane Katrina, 7, 19, 34, 39, 294–95
Hut, Piet, 159
hyperinflation, 264–65

ice-nine, 31, 32, 145, 156
Ikle, Fred, 165
Immersion, Ingestion, and Inhalation (III)
    attacks, 176
impact of X-event, 7–9
impact time, 8–9, 63, 304n
income distribution in US, 279–80
income inequality, 143, 293–94
incompleteness, 53–54

Indiviglio, Daniel, 259–60
information overload, 292
intelligent agents, in agent-based simulations, 283–84
intelligent machines, 239–46
interest rates, 263–64, 268
International Monetary Fund (IMF), 252, 258
Internet: dependence on, 77–80; origin of, 70–71; size of, 76–77
Internet addresses, 68–71, 76
Internet failure, 68–91, 306n; adding it all up, 89–91; deliberate attacks, 81, 85–89; examples of, 71–76; systemic crashes, 80–85
Internet Protocol (IP), 68–71, 75, 87–89
intrasociety dilemma, 204–5
Iran, 116, 117; nuclear program, 86–87, 172–73, 177, 187–90; oil reserves, 185; Stuxnet virus, 86–87, 307n
Iraq, 187; oil reserves, 185
Iraq war, 4, 98, 185
Ireland, 265
I, Robot (movie), 245
iRobot, Inc., 243
irrational exuberance, 37, 300
isolationism, US, 134–35
Israel, 172–73, 174, 175, 177, 190, 227
Italy strike of 2007, 180–82

Jaffe, Robert, 159
Japan: deflationary depression, 266–68, 316n; earthquake and tsunami of 2011, 8, 12–13, 256, 297–98
Joy, Bill, 239, 315n
just in time (JIT), 181

Kahn, Herman, 163, 311n
Kaminsky, Dan, 68–70, 75, 87–89, 306n
Keeler, William Henry ("Wee Willie"), 7
Kelly, Walt, 302
Kennedy, John F., Jr., 154
Kennedy, Paul, 12, 137
Khodorkovsky, Mikhail, 126
Kitty Hawk, USS, 165
Koo, Richard, 267, 316n
Kopp, Carlo, 113
Krakatoa eruption of 1883, 20–21, 61
Krioukov, Dmitri, 81
Krugman, Paul, 258–59, 315n

Kubrick, Stanley, 163
Kurzweil, Ray, 235–37, 248–49
Kuwait, 185, 187–88

Langmuir, Irving, 31
Large Hadron Collider (LHC), 147–51, 155–59, 311n
law of accelerating returns, 237, 241, 249
law of requisite complexity, 12, 304n
law of requisite variety, 56, 304n
Law of the Few, 200–201
laws of robotics, 244–45, 314n
Lehman Brothers, 38, 43
LePore, Theresa, 55
Libya, 9–13
Lights Out (Makansi), 222, 224
Limited Test Ban Treaty (LTBT), 167
limited war, 172–73, 177–78
liquefied natural gas (LNG), 224, 226
liquidity trap, 264
Little Ice Age, 170
Litvinenko, Alexander, 176
Lloyd, Seth, 47–48, 235
localization, 125, 128
Loescher, Michael, 132, 133
Lofgren, Eric, 208–10
Lokela, Mabako, 199
Loma Prieta earthquake of 1989, 296
"long tail" concept, 36
Lorenz, Ed, 54–55
Los Alamos National Laboratory, 145–46
Lucas, Robert, 255, 315n

McCarthy, Cormac, 115
McIlhenny's Tabasco Sauce, 50–51
MAD (Mutual Assured Destruction), 116–17, 164–65, 177, 311n
magnetic monopoles, 156
magneto-hydrodynamic device (MHD), 112–13, 116, 121
Maiello, Michael, 124
Makansi, Jason, 222, 224, 313n
malicious Internet attacks, 85–89
malware, 88–89, 306n
Manhattan Project, 145–46
Manning, Peyton, 284
manufacturing, 49–50
Mariner 9, 168
Marshall Plan, 9
Marx, Karl, 294

mathematical tools for anticipation, 278–80, 316n

Matrix, The (movie), 221–22

Matsuda, Hisakazu, 267

Mbeki, Thabo, 217–18

Mead, Carver, 233–34

Medvedev, Dmitri, 126

Merton, Robert, 37

Microsoft Windows, 25, 42–43

Mikko, Madis, 73

Militant Shangri-La scenario, 133

mini (micro) black holes, 149–52, 157–58

Minority Report (movie), 135

Mirkovic, Jelena, 88

Monod, Jacques, 40

Moore, Gordon, 233–35

Moore's law, 233–35, 237, 249

moral hazard, 255

Morgenstern, Oskar, 163–64

mousepox, 30

Mubarak, Hosni, 54, 74–75, 86

multitasking, 292

Murder in the Cathedral (Eliot), 257–58

Murphy's law, 235

nanobots, 32

nanotechnology, 31–32, 238

Nash, Jordan, 149

natural disasters, overview, 20–22

Navy, US, 165, 166

near-earth object (NEO), 21

Neuromancer (Gibson), 96

New York City blackout of 1977, 212–13, 215

New York power-grid collapse of 2003, 215–16

New York Stock Exchange (NYSE), 252

Nezami, Nashriyeh-e Siasi, 118

"nice" X-events, 9

Nikkei Stock Index, 266

Nineteen Eighty-Four (Orwell), 96

Nishioka, Junko, 268

No Free Lunch, 50–51, 301

normal probability distribution, 34–37, 35n

normal regime, 14–15

North American Electric Reliability Council (NERC), 215, 216

North Korea, 115, 117, 176

nuclear landscape, 161–79, 311n; adding it all up, 178–79; Cold War mental-ity, 161–66; damage-limitation ideas, 176–77; EMP and, 109–10, 112, 115, 117; history of technology, 145–46, 163–66; Iran program, 86–87, 172–73, 177, 187–90; "limited" nuclear war scenarios, 172–76; power grid vulner-abilities and, 224

nuclear terrorism, 23, 115–17, 162–63, 175–76; potential scenarios, 171–72, 175–76, 188–89, 311–12n

nuclear winter, 166–70, 311n

Occupy Wall Street, 28–29

oil supplies, 180–94, 312n; adding it all up, 191–94; global food supply and, 104; Italy strike of 2007, 180–82; peak oil question, 182–86, 191, 192–93; pos-sible scenarios, 186–91

On Thermonuclear War (Kahn), 163, 311n

Oppenheimer, J. Robert, 145–46

optical fiber cables, 83

ordinary and the surprising, 4–5

Orwell, George, 96

outsourcing energy sources, 224

Owens, R. C., 284

Ozzie, Ray, 42

Pakistan, 170, 173, 175

pandemics, 28, 195–211, 312–13n; adding it all up, 210–11; early-warning signs, 202–3; personal freedoms vs. public health, 203–6; short list of, 197–98; terminology, 197; three network principles, 199–202; three ways to stop, 207–8; virtual world simulation, 208–10

parallel vs. serial, 242

particle accelerators, 146–59, 310n

peak oil, 182–86, 191, 192–93, 312n

Persian Gulf, 186, 187–91

Pesek, William, 297–98

pesticides, 93, 100–101

Peter Lake experiment, 276–78

phosphorus, 92

physics, 144–60

phytophthora ramorum (PR) fungus, 94–95, 97, 102

Plague, The (Camus), 195–97, 312n

Plague of Justinian, 197–98

Planck, Max, 150

plant viruses, 94–98
Pollack, James B., 168–70
pollution, 226
population growth, and global food supply, 105, 106
Portugal, 265
poverty, 92, 124–25
power consumption, Internet, 82–83
power grid failures. *See* electric power grid failures
Power of Context, 201
preemptive nuclear strikes, 174–75
presidential election of 2000, 55
Pringle, David, 95–96
probability curve, 34–37, 35*n*
process innovation, 50–51
product diversification, 51
Project Proteus, 132–35, 310*n*
Project Starfish Prime, 111–12, 118
punctuated equilibrium, 137–38
Putin, Vladimir, 125–27, 309*n*

quantum vacuum collapse, 156–57
quarantine, 201, 204, 209–10
quarks, 151, 156

Rajaratnam, Raj, 258
rarity of X-events, 13, 38–39
Reagan, Ronald, 177
Red Capitalism (Walter and Howie), 52
Red Queen Hypothesis, 49–50
redundancy, 46, 301
Rees, Martin, 159, 303*n*
regional nuclear attacks, 175
Relativistic Heavy Ion Collider (RHIC), 153–55
Replogle, John, 100
resilience, 300–301
retail commerce, and Internet dependence, 79
retaliatory nuclear strikes, 175
Richardson, Bill, 224, 226
Road, The (McCarthy), 115
Robby, the Robot, 243–44
robotics, 238, 244–46, 314*n*
Roman Empire, 12, 44, 54, 139
Roomba, 243
router scalability, 84–85, 306–7*n*
runaway black holes, 156
Rushlo, Ben, 87–88

Russia, 88, 125–27, 140–41, 173

Sagan, Carl, 168–70
Samuelson, Robert, 258–59
San Diego blackout of 2011, 220–21
SARS (severe acute respiratory syndrome), 61, 189, 200, 206
Saudi Arabia, 92, 104, 185, 187–88
Saul, John Ralston, 124–25, 309*n*
scale-free network, 225
Schneider, Stephen H., 169–70
Schoch-Spana, Monica, 295–96
Scholes, Myron, 37
Schuchard, Max, 84–85, 307*n*
Schumpeter, Joseph, 269
Schwartz, Peter, 135–37
Science Fiction: The 100 Best Novels (Pringle, ed.), 95–96
Scud missiles, 115–16
Seabright, Paul, 254–55, 315*n*
second law of robotics, 244–45, 314*n*
seed banks, 97–98, 102
Sen, Amartya, 105
serial vs. parallel, 242
Shell Oil, 182
Silent Spring (Carson), 100
Singularity, 235–36, 237, 247–50, 314*n*
Singularity Is Near (Kurzweil), 235–37, 248–49
smallpox, 30, 197
Smith, Gary, 110
smoking, 203–4
social networking, 11–12
software failures, 81–85
soil erosion, 104, 106
Somalia, 228
South Africa, 217–18
Southwest blackout of 2011, 220–21
Soviet Union, 125–26, 136, 139, 161–66. *See also* Russia
Space Defense Initiative (SDI), 177
space defense systems, 177
Spain, 131, 265
Spanish flu outbreak of 1918, 30, 62, 198
Spears, Tara, 149
speculative bubbles, 259, 263, 266
Spengler, Oswald, 137
Spielberg, Steven, 135
Standard Model, 147–48, 160
Stickiness Factor, 201

Stiglitz, Joseph, 293–94, 317*n*
stock market, 36–37
Stoltenberg, Jens, 97
storage of electricity, 225
stories explaining X-events, 39
Strait of Hormuz, 187–88
strangelets, 150–51, 156, 159
Strategic Petroleum Reserve, 190
string theory, 148–50
Strock, Carl, 34–35, 305*n*
Strogatz, Stephen, 6, 14
Stuxnet virus, 86–87, 307*n*
subsistence diets, 105
Suez Canal, 83
Sumatra eruption of 1883, 20–21, 61
Sun Tsu-yun, 117–18
superbugs, 206
superspreaders, 200–201
superstring theory, 149–50
supersymmetry, 149–50
survivalism, 141, 244
sustainability, 15, 46, 274
Svalbard Global Seed Vault, 97–98, 102
Sweden, 71–72
Syria, 9, 12, 174
systemic Internet crashes, 80–85
systemic traits, 48–49
Szekely, Louis ("Louis CK"), 77

Tainter, Joseph, 44, 45–46, 49, 305*n*
Taiwan earthquake of 2006, 83
Takemori, Shumpei, 267–68
Taleb, Nassim Nicholas, 37–40, 54, 305*n*
Tamiflu, 207–8
taxes (tax policy), 56, 143, 260, 265
technology, 233–50; GNR problem, 237–39, 247; human brain vs. computer, 240–43; Moore's law, 233–35, 237; Singularity, 235–36, 247–50
Teller, Edward, 145–46
Tesla, Nikola, 222–23
Tevatron, 149, 154
Texas City Refinery explosion of 2005, 292
Texas drought of 2011, 228
Thailand floods of 2011, 294–95
theory of everything (TOE), 147
thermodynamics, 8, 245
third law of robotics, 244–45, 314*n*
Thomas, Tom, 132, 133
Thompson, Derek, 259–60

Three Mile Island accident, 290
Through the Looking Glass (Carroll), 49
timescale, 66
Tipping Point, The (Gladwell), 200–201, 313*n*
Toba eruption, 20–21, 304*n*
Toon, O. Brian, 168–70
total impact, 8–9
Toynbee, Arnold, 137
transportation system, and Internet dependence, 80
Treasury Department, US, 87, 251, 253, 260
Trinity (nuclear test), 145–46
trust, in financial system, 254, 258–59
tuberculosis, 203, 205
Tunisia, 27, 54, 129
Turco, Richard, 168–70
Turing, Alan, 239
Turing test, 239–40
Twitter, 11–12, 82, 88
typhoid fever, 204
Typhoid Mary (Mary Mallon), 204–5

Ugly Swan Paradox, 298
Ulam, Stan, 236
undecidability, 53–54
underground bunkers, 140–41
unemployment, 142–43, 259
unfolding time, 8–9, 63, 304*n*
United Arab Emirates, 185, 187
United Nations (UN), 136, 229
Until the End of the World (movie), 109–10
Upside of Down, The (Homer-Dixon), 45

vaccinations, 207
Vanishing of the Bees (movie), 98–99
Van Valen, Leigh, 49
"variety," 56
varroa mites, 101
Venezuela, 188–90
Viagra, 11
Vickers, Damon, 251–52, 253
Vinge, Vernor, 235–37, 247–48, 314*n*
virtual water, 230, 314*n*
viruses: bees, 101; Internet, 88–89; plant, 94–98
Volkova, Lyubov, 126, 127
Vonnegut, Kurt, 31, 32, 145, 156
Von Neumann, John, 163–64, 236

se

Wagner, Walter, 154, 155–57
Waldcock, Bill, 288
Wall Street Crash of 1929, 266
Walter, Carl E., 52
water supply, 104, 106, 227–32, 313–14n
Watts, Duncan, 202
weather forecasting, 5, 7, 169
Weldon, Curt, 110
Wenders, Wim, 109–10
Westinghouse, George, 223
wheat rust, 93
white dove, 40–41
Wilczek, Frank, 154, 310n
"willful blindness," 291–92, 317n
Wolfe, Nathan, 198–99, 312–13n

Wood, Lowell, 110, 118
World Health Organization (WHO), 202, 205
World Is Flat, The (Friedman), 122, 124
World of Warcraft, 208–10, 313n
worms, Internet, 88–89

X-events regime, 14–15
X-risk, 299–301

Yankee Going Home scenario, 134–35
Yellowstone National Park, 64
Yeltsin, Boris, 125–26

zeroth law, 245